RUNNING OUT OF TIME
WILDFIRES AND OUR IMPERILED FORESTS

amplifypublishinggroup.com

Running Out of Time: Wildfires and Our Imperiled Forests

©2023 David L. Auchterlonie, Jeffrey A. Lehman, Crowbar Research Insights LLC. All Rights Reserved. No part of this publication may be reproduced, stored in a retrieval system or transmitted in any form by any means electronic, mechanical, or photocopying, recording or otherwise without the permission of the authors.

The views and opinions expressed in this book are solely those of the authors. These views and opinions do not necessarily represent those of the publisher or staff. The advice and strategies found within may not be suitable for every situation. This work is sold with the understanding that neither the authors nor the publisher are held responsible for the results accrued from the advice in this book.

Neither author, nor Crowbar Research Insights, LLC itself have any conflicts of interest with any third party regarding this book. The research, findings, and recommendations are our own without influence or agenda from any third party.

Cover Photo Credit: Photograph by Noah Berger/AP Photo, August 18, 2020

For more information, please contact:
Amplify Publishing, an imprint of Amplify Publishing Group
620 Herndon Parkway, Suite 220
Herndon, VA 20170
info@amplifypublishing.com

Library of Congress Control Number: 2023908736

CPSIA Code: PRV0423A

Hardcover ISBN: 978-1-63755-783-9
Paperback ISBN: 978-1-63755-874-4
eBook ISBN: 978-1-63755-784-6
Running Out of Time is also available on audiobook.

Printed in the United States

We dedicate this book to each of the courageous local, state, and federal firefighters, hotshot crews, mechanized equipment operators, smoke-jumpers, aerial firefighting crews, first responders, and supporting teams who keep us safe from wildfires.

We also acknowledge the thoughtful stewards of forests worldwide, especially the Indigenous peoples, who practice forest maintenance crafted through thousands of years of experience. We have so much to learn from them.

COVER PHOTO
PHOTOGRAPHER'S NOTE

The image of flames cresting a hill behind a flag marks an especially intense moment during a vigorous streak of fire coverage. Driven by strong winds and exceptionally dry vegetation following a heat wave, the LNU Lightning Complex exploded on August 18, 2020. The flagpole sits at a crossroads near Lake Berryessa. With the fire rapidly advancing and about to cut off Highway 128, I had to decide whether to take to road to Spanish Flat, which would experience intense fire behavior with basically no exit routes or play it safer and stay on Highway 128. I decided to ride out the fire in Spanish Flat, photographing as the fire destroyed much of the community. I ended up working until 4 am that day, by which time the fire had reached Vacaville, leveling hundreds of homes and claiming multiple lives.

—Noah Berger

RUNNING OUT OF TIME

WILDFIRES AND OUR IMPERILED FORESTS

DAVID L. AUCHTERLONIE
AND
JEFFREY A. LEHMAN

CONTENTS

Foreword 1

Introduction 3

Acronyms 8

CHAPTER 1 The Impact of Wildfires 13

CHAPTER 2 The Role of Federal and State Agencies 29

CHAPTER 3 Cost of Wildfires and Forestland Management . 51

CHAPTER 4 Wildfire Occurrences and Responses 69

CHAPTER 5 Personnel and Mechanized Assets
for Fighting Wildfires 105

CHAPTER 6 Aviation Assets to Contain and Extinguish
Wildfires 123

CHAPTER 7 Costs and Benefits of
Timely Deployment of Assets 157

CHAPTER 8 International Wildfire and
Forestland Management Observations 183

CHAPTER 9 Review of Forestland Management Practices . 207

CHAPTER 10 Forest Carbon Storage,
Carbon Emissions, and Carbon Offsets 247

CHAPTER 11 Put the Fires Out!
Analysis and Recommendations 273

Conclusion 309

Appendix 315

Endnotes 339

Bibliography 383

Acknowledgments 397

About the Authors 401

FOREWORD

Neglected for decades, America's public forestlands are highly vulnerable to increased catastrophic wildfires, insects, and diseases that threaten the health of the nation's private forestlands as well. Communities, wildlife, water resources, and natural landscapes face increased jeopardy without additional human and capital resources. At least 100 million acres of national forests are at risk of severe wildfire without active forest management.

Unfortunately, our communities have suffered from the lack of effective forest management, wildfire suppression, and timber harvests. Today anti-forestry obstruction and litigation prevent public land managers from implementing forest projects that reduce the risks of wildfires and the carcinogenic smoke they create. Regardless of the misguided good intentions behind these efforts, they are literally loving the forests they wish to protect to death.

This book thoroughly examines how our federal and state firefighting agencies lack the funding to meet today's wildland fire crisis. It provides insights into current firefighting and forest management practices and proposes solutions.

The authors suggest using novel private-public partnerships to muster the resources needed to act now to implement practical and necessary solutions as were accomplished in Texas. The book offers the opportunity to time-test scientifically proven alternatives and emerging technological advancements to determine the best forest management practices based on regional forest conditions. The incentives identified also allow ample investment to address our neglected forest management and wildfire suppression needs.

Tom Boggus
Chairman, National Association of State Foresters Foundation
Director Emeritus, Texas A&M Forest Service (2008—2021)

INTRODUCTION

The authors of *Running Out of Time: Wildfires and Our Imperiled Forests* have more than 110 years of combined experience in both private and public sectors, but experience alone is not what motivated us to write this book. This project is the product of our passion and thirst for knowledge and solutions to complex problems that have a long-standing, long-term devastating impact on so many individuals and society.

DAVID L. AUCHTERLONIE

I grew up and still reside in Southern California. During the 1950s and 1960s, I observed the transformation of the San Fernando Valley from vast orchards into a growing residential suburb of Los Angeles, with housing tracts, shopping malls, and freeways replacing historic orange groves. Wildfires were an annual event in the nearby Angeles Crest Forest, Santa Monica Mountains, and San Gabriel Mountains.

Our family vacations always included camping trips to Sequoia National Park, Kings Canyon National Park, and Yosemite National Park, a tradition I continued by backpacking and skiing the Sierras, the Cascades, and the Rocky Mountains with my three sons. Now mule trips replace backpacking to enjoy the majestic Sierras and rejuvenate my soul. These pilgrimages reconnect me and add to the lasting memories of past trips. Books by John Muir, Henry David Thoreau, Dee Brown, Edward Abbey, Jack London, David McCullough, and others influenced my appreciation of nature and the West.

As the eldest of nine children, I grew up in a family in which chaos and order were constant companions, despite my parents' best efforts. This upbringing influenced my career decisions as well. I became a corporate turnaround manager for clients who were either financially distressed, mismanaged, or lacked a clear strategy to serve customers.

Through The Scotland Group, Inc., a boutique turnaround management firm I founded in 1986, my firm successfully managed over 200 client companies through complex financial and operating crises. Enacting change out of chaos required my development of hands-on operating tactics to implement the successful turnaround strategy I developed for these clients.

In 2012 my colleague (and coauthor) and I attempted to acquire an aerial firefighting company in the belief that the USDA Forest Service would welcome additional resources to assist its fight against increasing wildfire risk. Unfortunately, we learned the Forest Service had other agendas, which influenced the recommendations contained in this book.

JEFFREY A. LEHMAN

My involvement with forests and timberlands began as a Boy Scout in a suburban county north of New York City, during which time I learned about forests, respecting the woods, and caring for them. At the same time, I developed a fascination for flight and airplanes, which first started when I sat in the back of a Piper Cub when I was eight years old.

In my early twenties, I was promoted to a newly created position in a European headquarters of a major logistics company in Rotterdam, Netherlands, to help manage offices in Europe, the UK and Ireland, Africa, the Middle East, and India. During this assignment, I traveled extensively throughout those areas and spent many weekends, among other things, walking through the local woods. I vividly remember the trees as different species from anything I had ever seen and called those walks "enlightening—never to be forgotten."

Later in life, I met with a venture capitalist who purchased and ran a company headquartered in Maine that converted company-owned pine trees into matches and toothpicks. It was another lesson about forests, but this time about economics.

After many years in air cargo market research and development and logistics, I started a consultancy and took on an assignment for McDonnell Douglas and later Boeing. During that period, I worked on a number of projects, one of which focused on the C-17 and its operational impacts on global logistics. This effort also included a broader assessment of the C-17 as a commercial variant that could be used in a number of multi-mission roles across several industry groups and government.

Through another client, an airborne firefighting company, I became intimately familiar with the Forest Service's aircraft firefighting capabilities—an experience that resulted in my deep interests in Forest Service logistics, aerial firefighting, fleet planning for aerial firefighting, and the finances required for success.

A PERSONAL NOTE FROM THE AUTHORS

We discovered this story required an analysis of forest management. More than 100 million acres of poorly managed federal forests are catalysts for intense and devasting wildfires. This book is our focus on government bureaucracy's neglect of our forests for more than a century . . . and counting. It describes wildfire suppression and forest management, including the agencies responsible, costs involved, assets utilized, and practices employed. It also includes global information for comparison to the worldwide challenge of forest management and wildfire suppression. Further, we explore increased global temperature and the task involving carbon dioxide sequestration linked to forests.

Finally, we analyze the current situation in the U.S. and propose realistic, incentive-driven solutions to fund the significant costs needed to restore our neglected forests. Annually, the cost of this neglect in the form of devasting wildfires exceeds $300 billion. Toxic carcinogenic wildfire smoke impacts two-thirds of the U.S. population. National and state park closures significantly impact recreational use. Watershed and wildlife suffer severe loss, all of which is unacceptable.

Our blueprint complements the already extensive research performed by academics, forestry experts, and government agencies to help achieve a reduced number and ferocity of wildfires, a lower volume of toxic carcinogenic smoke, and improve the health of our forests.

We recognize the excellent effort of the federal, state, and local employees who do the work to preserve our forests and prevent and suppress wildfires. There are many dedicated people performing unheralded tasks. Unfortunately, the same cannot be said about the leadership in many of the bureaucratic agencies.

We acknowledge the sources accessed to create this book—the acknowledgments, bibliography, and footnotes to each chapter contain references and secondary sources. We conducted interviews with personnel inside and outside government agencies and list some in the acknowledgments section. Other sources provided invaluable information based on firsthand experiences. Because many of these individuals still work for or with various government agencies, their names have not been published to avoid the risk of reprisal.

Finally, we do not hold ourselves out to be forestry experts. Our proposed solutions rely on experts in each of their respective fields for implementation. In the end, we are turnaround, strategic planning, organization, and aviation experts. The glaring shortcomings we describe in this book require commonsense business solutions. We also believe incentive-based capitalism can play a significant role in providing the funding necessary. We hope our findings and recommendations provide the catalyst desperately needed to bring order to a mismanaged situation.

ACRONYMS

AFUE	Aerial Firefighting Use and Effectiveness Report
BIA	Bureau of Indian Affairs
BIFC	Boise Interagency Fire Center
BLM	Bureau of Land Management
CAL FIRE	California Department of Forestry and Fire Protection
CAMS	Copernicus Atmosphere Monitoring Service
CCS	Carbon Capture and Storage
CH_4	Methane
CO_2	Carbon Dioxide
CO_{2e}	Carbon Dioxide Equivalent
CRS	Congressional Research Service
CWN	Call When Needed
DNR	Department of Natural Resources
DOI	United States Department of the Interior
EPA	United States Environmental Protection Agency
ESG	Environmental, Social and Governance
EU	Essential Use
FEMA	United States Federal Emergency Management Agency
FIRMS	NASA Fire Information for Resource Management System

ACRONYMS

FLAME	Federal Land Assistance, Management, and Enhancement Act
FPM	Forest Production Modernization Strategic Framework for Long-Term Action
FPZ	Forest Protection Zones
FS	United States Department of Agriculture Forest Service
FWS	United States Fish and Wildlife Service
FY	Fiscal Year
GACC	Geographical Area Coordination Center
GAO	United States Government Accountability Office
GHG	Atmospheric Greenhouse Gas/Greenhouse Gas
GNA	Good Neighbor Authority
GOES	Geostationary Operational Environmental Satellites
IC	Incident Commander
ICS	Incident Command System
LATs	Large Airtankers
MAFFS	Modular Airborne Fire Fighting Systems
MAP	Mandatory Available Period
MMTs	Million Metric Tons
MOU	Memorandum of Understanding
N_2O	Nitrous Oxide
NAFC	National Aerial Firefighting Centre

NASF	National Association of State Foresters
NASA	National Aeronautics and Space Administration
NCEI	NOAA National Centers for Environmental Information
NEPA	National Environmental Policy Act
NFPA	National Fire Protection Association
NICC	National Interagency Coordination Center
NIFC	National Interagency Fire Center
NIFRMA	National Indian Forest Resources Management Act
NIMS	National Incident Management System
NIST	National Institute of Standards and Technology
NMAC	National Multi-Agency Coordinating Group
NOAA	National Oceanic and Atmospheric Administration
NPS	National Park Service
NRF	National Response Framework
NSC	National Strategic Committee
NWCG	National Wildfire Coordinating Group
NWS	National Weather Service
NYDF	New York Declaration on Forests
PCWA	Placer County Water Agency
PG&E	Pacific Gas and Electric
PM	Particulate Matter

ACRONYMS

$PM_{2.5}$	Fine Particulate Matter
PNAS	Proceedings for the National Academy of Sciences
PPM	Parts Per Million
RAWS	Remote Automated Weather Stations
RDS	Retardant Drop Systems
REPLANT	Repairing Existing Public Land by Adding Necessary Trees Act
SEATs	Single-Engine Airtankers
UN	United Nations
USDA	United States Department of Agriculture
USFA	United States Fire Administration
UTF	Unable to Fill
VLATs	Very Large Airtankers
WFLC	Wildland Fire Leadership Council
WHO	World Health Organization
WFM	Wildland Fire Management
WUI	Wildland-Urban Interface
WWF	World Wildlife Fund

CHAPTER 1

THE IMPACT OF WILDFIRES

Fire seasons now average seven months each year, almost two and a half months longer than in 1970.[1]

The Earth is on fire. This is not hyperbole. Wildfires impacted every continent in 2020, with nearly 1 billion acres destroyed worldwide.[2,3]

In the United States, 10.1 million acres were burned by over 59,000 wildfires in 2020, equivalent in size to the states of Maryland and Delaware combined, according to the Congressional Research Service.[4] Bushfires in Australia burned an astounding 47 million acres in 2019 through 2020, equivalent in size to the entire state of Washington.[5]

In recent years, countless more areas across the globe experienced historically dangerous wildfires (see fig. 1).

Wildfires typically occur during periods of increased temperature and drought. They are anything but a new phenomenon. Archaeologists discovered the first wildfires occurred more than 7,000 years ago.[6] Scientists determined that Greeks, Romans, and early Chinese civilizations faced wildfire threats.

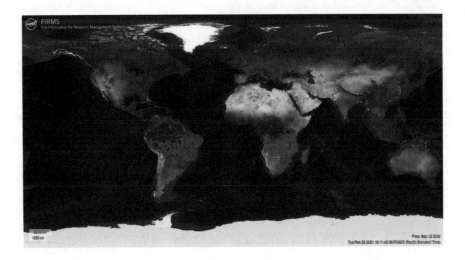

Figure 1. Global image of wildfires on September 12, 2020. (Figure by NASA Fire Information for Resource Management System (FIRMS), September 12, 2020, https://firms.modaps.eosdis.nasa.gov/map/#t:adv;d:2020-09-11..2020-09-12,2020-09-13;l:country-outline;@-24.4,9.4,3z).

NORTH AMERICA

Similarly, North America's Indigenous people lived with wildfires impacting their daily way of life, but over time, they applied the knowledge gained to maintain their ecosystems. Charcoal evidence from geological and archaeological research suggests that Native Americans experienced a general decrease in America's wildfires until about 1750, when Western culture expanded geographically. Geological data between 1750 and 1870 implies a period of increased fire frequency in North America attributed to human population growth and influences such as land-clearing practices. This period was followed by an overall decrease in burning in the twentieth century, linked to the expansion of agriculture, increased livestock grazing, and fire prevention efforts.[7]

Nonetheless, as the population settled into wildland-urban interface (WUI) areas, wildfires again increased, particularly in the past fifty years. Human activity in WUI areas brings not only exposure of life and property to wildfires but an additional source of ignition. Sparks from machinery, trains, escaping campfires, poorly planned agricultural and forestland management fires, arcing from electrical utility lines, and unwatched campfires are all human causes of wildfires; however, more than 60 percent of all wildfires in the United States in the past twenty-five years continue to be caused by Mother Nature as the result of lightning strikes igniting accumulated undergrowth in our forests (see fig. 2).

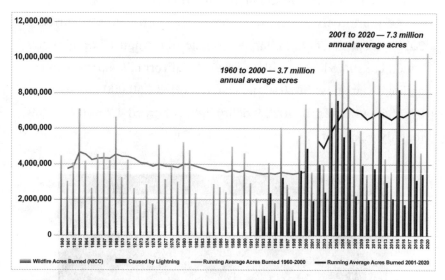

Figure 2. United States acres consumed by wildfires—1960-2020. Annual wildfire acres burned 1960-2020 and lightning-caused acres burned 1992-2020, with running annual average 1960-2020 and 1992-2020. (Lightning-caused fire reporting began in 2001). (Figure created by Crowbar Research Insights LLC™ with data from the National Interagency Coordination Center, https://www.nifc.gov/. © Crowbar Research Insights LLC™. All rights reserved).

Acreage burned in the United States increased from an average of 3.3 million acres in the 1990s, roughly the size of Connecticut, to an annual average of 7 million acres since 2000, about the size of

Massachusetts. On average, 3.3 million acres burned in the 1990s. This number rose to 10.1 million acres in 2020, a 206 percent increase.[8] But there were more wildfire incidents in 1991 (75,754) as compared to 2020 (58,258), indicating the size and severity of wildfires is increasing.[9]

The largest amount of acreage burned (2.4 million) in 2019 was in Alaska, but California had the most wildfires.[10] Three of Colorado's largest wildfires in its history occurred in 2020.[11] Colorado's catastrophes consumed 625,000 acres, about six times the size of Denver, and cost more than $200 million in fire-suppression costs.[12]

Of the top six largest wildfires in California, five occurred in 2020—most notably, the largest to date, the August Complex fire, which consumed 1.03 million acres.[13] California's numbers mirror national statistics. In 2019 the state reported 260,000 acres burned by 7,860 wildfires. In 2020 9,639 wildfires burned 4.2 million acres, roughly 4 percent of California (see fig. 3).[14]

Figure 3. A home is engulfed in flames during the Creek Fire in the Tollhouse area of unincorporated Fresno County, California, early on September 8, 2020. (Photograph by Josh Edelson/AFP via Getty Images, September 8, 2020, https://www.gettyimages.com/detail/news-photo/home-is-engulfed-in-flames-during-the-creek-fire-in-the-news-photo/1228399068.)

In early October 2020, sixty-five large fires consumed more than 2 million acres in California, Idaho, Montana, Oregon, Washington, and five other states. In Oregon, an estimated 4,000 homes were damaged or destroyed by wildfires, forcing the evacuation of thousands to escape the flames that scorched more than 230,000 acres—an area fourteen times the size of Portland, Oregon. In California, fires stretching across approximately 800 miles of landscape burned from the north all the way to the Mexican border (see fig. 4). In Washington, more acres were burned in 2020 than in the past twelve fire seasons combined.[15] The evidence is clear: wildfires are increasing in number and intensity.

Figure 4. A law enforcement officer watches flames launch into the air as fire continues to spread at the Bear Fire in Oroville, California, on September 9, 2020. (Photograph by Josh Edelson/AFP via Getty Images, September 9, 2020, https://www.gettyimages.com/detail/news-photo/law-enforcement-officer-watches-flames-launch-into-the-air-news-photo/1228424195.)

WHY WILDFIRES ARE INCREASING

Many factors, including accumulation of forest undergrowth, insect infestation, and WUI, contribute to hotter, drier, and longer wildfire seasons. Just as telling, research data indicates that from 1970 the annual average temperatures in the western United States increased 1.9 degrees Fahrenheit.[16] Winter snow is melting earlier than in previous decades, leading to drier forests, while large areas of dead trees increase the likelihood of wildfires, possibly due to uninhibited infestation by insects, including the bark beetle.[17]

Halofsky, Peterson, and Harvey, in *Fire Ecology*, dramatically warned of the ramifications. "Changing climate and fire frequency, extent, and severity are likely to influence forest regeneration processes, thus affecting the structural and compositional trajectories of forest ecosystems," they wrote. "[W]armer weather is expected to affect regeneration through increased fire frequency. As fire-free intervals shorten, the time available for plants to mature and produce seed before the next fire will be limited. Such changes in fire-free intervals can have significant effects on postfire regeneration because different plants have varied adaptations to fire. Species that resprout following fire may decline in density, but species that are fire-killed and thus require reproduction from seed may be locally eliminated."[18]

Research by the University of Oregon concludes reforestation following a large wildfire event may take between fifteen and twenty years.[19] Tree planting helps restore these areas to become forested once again.[20]

Forests cover 31 percent of the Earth's surface.[21] Lightning is the cause of most naturally occurring wildfires, which tend to burn more acreage than those started by humans; however, it is humans who cause most wildfire incidents (approximately 89 percent).[22] While the larger wildfires occur primarily in the West, more wildfires occur east of the Mississippi.[23] Population growth continues to encroach on forests due to WUI, increasing the likelihood of more damaging human-caused wildfires through burning debris such as dead vegetation, unattended campfires, equipment sparks, arcing power lines, and discarded cigarettes.[24] When forests are purposefully intruded upon for new homes and communities—a prime example of WUI—fire risk increases dramatically. It obviously is more difficult to let natural wildfires burn in WUI areas, which are the fastest-growing land-use type in the contiguous United States.[25] Most new WUIs are the result of new housing (97 percent). According to the U.S. Department of Agriculture, there are more than 70,000 communities and 46 million homes at risk from wildfires (see fig. 5).[26]

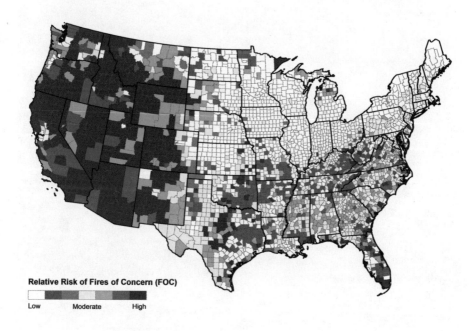

Figure 5. Areas in the United States at risk for large, long-duration wildfires. (Figure by the Science Analysis of the National Cohesive Wildland Fire Management Strategy, 2022, https://cohesivefire.nemac.org/option/10.) [Figure credit: NEMAC and the Eastern Forest Environmental Threat Assessment Center (EFETAC)]

AIR QUALITY AND WILDFIRES

In addition to the direct impacts, there are other substantial effects on society due to fires. Smoke from wildfires impacts large population centers, some of them thousands of miles from the original wildfire site. Recreation centers and national parks frequently are closed in many states during August and September, as was the case in 2020, because of toxic air quality conditions.[27] The expansive nature of large fires and the smoke they create impacts inhabitants living on half the land mass of the United States, or an estimated 212 million people.[28] Figures 6 and 7 demonstrate examples recorded by National Aeronautics and Space Administration (NASA) satellites of the extensive residual smoke impact from fires in California and Colorado in 2020 (see figs. 6 and 7).

THE IMPACT OF WILDFIRES

Figure 6. Yosemite National Park, Half Dome September 6, 2020. Smoke and ash from the Creek Fire, which began on September 4, 2020, in the Sierra National Forest. On September 7, 2020, the USDA Forest Service closed all Southern California National Forests due to unprecedented fire conditions. (Photograph by the *Fresno Bee*, September 8, 2020, https://www.fresnobee.com/news/nation-world/national/article245559400.html). [Photo credit: From the *New York Times* © 2020 the New York Times Company. All rights reserved. Used under license.]

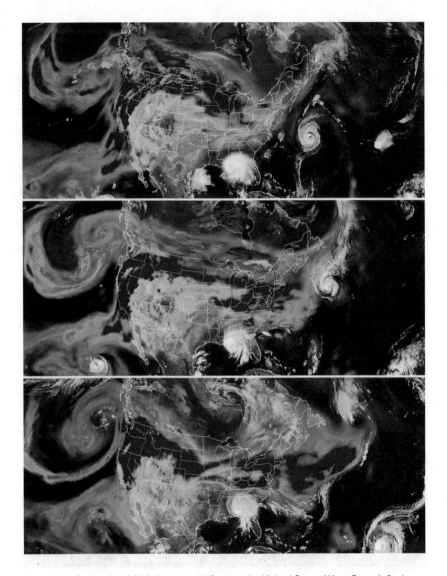

Figure 7. In September 2020, historic wildfires on the United States West Coast lofted plumes of smoke high into the atmosphere. Pushed by prevailing winds that sweep air from west to east, satellites tracked the smoke as it spread widely across much of the continental United States. A second hazard—tropical cyclones—also helped steer the high-flying smoke plumes as they streamed over the Midwest and Northeast between September 14 and 16, 2020.

The series of images above shows the abundance and distribution of black carbon, a type of aerosol found in wildfire smoke, as it rode jet stream winds across the United

States. The black carbon data comes from the GEOS forward processing (GEOS-FP) model, which assimilates information from satellite, aircraft, and ground-based observing systems. The Visible Infrared Imaging Radiometer Suite (VIIRS) on the NOAA-NASA Suomi NPP satellite acquired the images of the storms. (Photography by the Earth Observatory, September 14–16, 2020, https://earthobservatory.nasa.gov/images/147293/a-meeting-of-smoke-and-storms).

Exceptionally fine particulate matter ($PM_{2.5}$) is carried within the smoke and ash. It is small enough to penetrate deep into the lungs and cross into the bloodstream. The World Health Organization (WHO) has determined that $PM_{2.5}$ causes acute respiratory issues, such as asthma, and is increasingly linked to death from heart and lung disease (see fig. 8).[29]

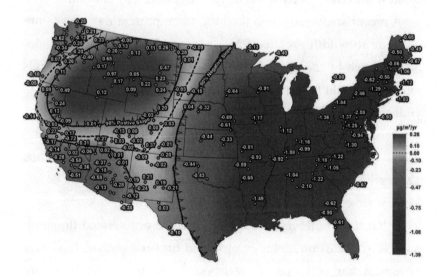

Figure 8. Fine particulate matter trends from 1988–2016. Wildfires are a major source of fine particulate matter (diameter <2.5 μm; $PM_{2.5}$), which is a health hazard. Since the mid-1980s, the total U.S. area burned by wildfires has been increasing, with fires in the Northwest United States accounting for ~50–60 percent of that increase. In the Northwest United States, a positive trend in the 98th quantile of $PM_{2.5}$ in contrast to other areas of the country. (Figure by the Proceedings for the National Academy of Sciences [PNAS], July 31, 2018, U.S. particulate matter air quality improves except in wildfire-prone areas | PNAS).

The Copernicus Atmosphere Monitoring Service (CAMS) monitors emissions from global fires and predicts how these pollutants will affect global air quality. In 2019 CAMS tracked emissions and activity of more than 100 wildfires in the Artic Circle and Alaska. This tracking revealed that in the first half of July, more than 31 megatons of CO_2 were released from these wildfires.[30] In the same year, CAMS monitored nearly 400 wildfires in Alaska. The smoke they emit poses a health risk to those in the state and much farther as these pollutants are blown thousands of miles, contributing to poor air quality globally.[31] In fact, using ozone observations, satellite data, and specific models, researchers attributed poor air quality in Houston, Texas, in July 2004 to forest fires that started a week prior in Alaska and Canada.[32]

A recent study estimated the long-term, present-day, and future exposure to wildfire-related $PM_{2.5}$ across Alaska. Researchers determined that by the mid-twenty-first century, nearly all of Alaska's population could be exposed to increases of 100 percent or more in wildfire-specific $PM_{2.5}$ levels.[33]

According to one study, deaths from wildfire smoke and $PM_{2.5}$ could double this century in the United States.[34] The outlook for Europe is also grim. Worldwide, the study estimates 340,000 premature deaths per year are caused by wildfire particulate matter.[35]

Other direct effects of wildfires include accelerated flooding, massive soil erosion, delay in crop and timber harvests, landmass movement, and pollution of water bodies. Not to be overlooked are additional social impacts, such as disruptions to road and air traffic, business closures, and loss of employment during and immediately after a wildfire. Further, long-term reductions of tourism, aesthetic values of the landscape, and home values are all adverse economic consequences of wildfires (see fig. 9). As a society, are we prepared for further disruptions and lockdowns as unhealthy air quality forces us to stay indoors during wildfire seasons each year?

Figure 9. Lefthand Canyon in Boulder County, after the floods of September 2013. (Photograph by *Boulder Daily Camera*, September 2013, https://www.weather.gov/bou/Number1September2013Floods). [Photo credit: MediaNews Group/Boulder Daily Camera via Getty Images.]

ENVIRONMENTAL IMPACT

Beyond the enormous quality of life issues and economic impacts, wildfires produce approximately 10 percent of the greenhouse gas emissions each year.[36,37] As wildfires grow in intensity and fire seasons extend, this annual percentage is expected to increase. In 2021 wildfires emitted an estimated 1.76 billion metric tons of carbon, according to CAMS.[38] Wildfire emissions are more than 50 percent of the global CO_2 emissions created by automobiles. Some environmentalists warn that large wildfires will become the norm because of global warming, thereby almost guaranteeing an increase in greenhouse gases for current levels (see fig. 10).

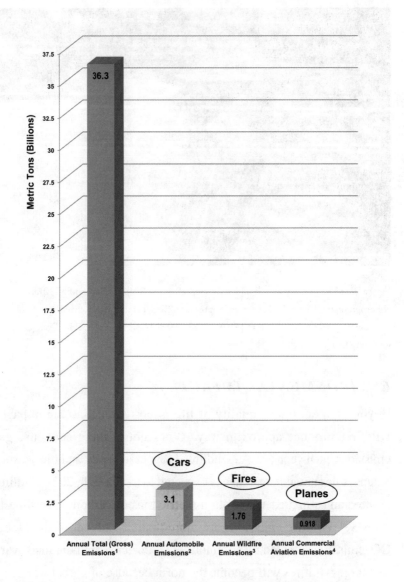

Figure 10. Selected drivers of global GHG emissions. (Figure by Crowbar Research Insights LLC™. Sources: (1) https://www.iea.org/reports/global-energy-review-co2-emissions-in-2021-2; (2) https://www.iea.org/topics/transport; (3) https://atmosphere.copernicus.eu/copernicus-2021-saw-widespread-wildfire-devastation-and-new-regional-emission-records-broken; (4) https://theicct.org/publication/co2-emissions-from-commercial-aviation-2013-2018-and-2019/, © Crowbar Research Insights LLC™. All rights reserved.)

Despite their severity, wildfires and their toxic transmissions are the most controllable of all greenhouse gas emissions. We can and must extinguish wildfires faster to *immediately* reduce greenhouse gas emissions and toxic $PM_{2.5}$ pollutants.

CAUSE FOR OPTIMISM

Despite the alarming trends described above, wildfire damages can be reduced and improved quality of life achieved through enhanced wildfire and forest management practices. This book provides a blueprint; however, major structural changes will be required, particularly to the governmental agencies responsible for our treasured forests and wildfire management.

THE KEY TAKEAWAYS FROM THIS CHAPTER ARE:

- Wildfires produce more than half the CO_2 emissions produced globally by automobiles.

- Globally, wildfires are increasing in number and intensity.

- The rapid expansion of wildland-urban interface contributes to the risk of wildfires.

- Wildfires pose a danger to public health because of their toxic air quality and greenhouse gas emissions.

CHAPTER 2

THE ROLE OF FEDERAL AND STATE AGENCIES

No government ever voluntarily reduces itself in size. Government programs, once launched, never disappear. Actually, a government bureau is the nearest thing to eternal life we'll ever see on this earth!

—*Ronald Reagan*[1]

A brief review of United States history can help explain the current state of forest management and wildfire prevention and containment. During the nineteenth century, United States' policy encouraged rapid settlement and economic development of its western territories. Among the approaches were the transfer of federal lands to individual farmers, ranchers, and corporations, especially the railroad companies that built the transportation infrastructure. After 1850, the U.S. population grew rapidly (20 to 25 percent per decade), and settlement of the western territories accelerated.[2]

That's when concerns first arose about environmental and economic implications of rapid development, including: (1) expedited deforestation (forests were cleared for agriculture at a rate of almost 9,000 acres per day); (2) massive wildfires from logging and land clearing (burning 20 to 50 million acres annually); (3) poor timber and forest management resulting in an estimated 80 million acres of "stump lands" (equivalent to the size of New Mexico), with little forest replenishment by 1920; (4) significant soil erosion in wildfire areas caused by wind and rain; and (5) significant wildlife depletion due to commercial hunting and subsistence use.[3,4]

FEDERAL AGENCIES

Policy and institutional changes in the early twentieth century were driven at the federal level instead of states or regions to address these societal anxieties of forest and wildfire management. As a result, centralized federal bureaucracies were created (see fig. 1).

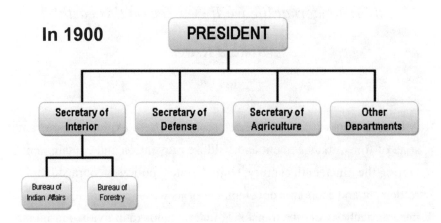

Figure 1. Organizational relationship of federal forest and wildfire management agencies in 1900. (Figure created by Crowbar Research Insights LLC™, © Crowbar Research Insights LLC™. All rights reserved).

Over the past 117 years, other agencies, bureaus, programs, centers, and councils were established within these centralized federal bureaucracies. In response, individual states began to develop their own forest management and natural resource departments as early as 1910. Unsurprisingly, forest and wildfire management are now composed of a multifaceted and complex federal, state, and local agency system.

At the federal level, responsibility for forest and wildfire management of 615 million acres of federal lands falls under the jurisdiction of these departments: Department of Agriculture, Department of the Interior (DOI), and Department of Homeland Security. Several agencies, bureaus, and councils are tasked with wildfire management within these departments.

U.S. DEPARTMENT OF AGRICULTURE

Under the U.S. Department of Agriculture (USDA), the agency responsible for forest and wildfire management is the USDA Forest Service (FS). Established in 1905, the FS mission is "to sustain the health, diversity, and productivity of the Nation's forests and grasslands to meet the needs of present and future generations." The FS oversees wildfire management for 193 million acres of the National Forest System. According to the USDA Budget Justification Fiscal Year (FY) 2021, wildland fire management is a top priority for the FS. In fact, instead of being a seasonal activity, the FS now responds to fires year-round, as the number increases every year (see fig. 2 and 3).[5-7]

Figure 2. The U.S. Department of Agriculture South Building located at 14th Street and Independence Avenue, S.W., in Washington, DC. (Permission from iStock)

Figure 3. The USDA Forest Service building in Washington, DC. (Photograph by USDA Forest Service, https://www.fs.fed.us/emc/.)

The fiscal year (FY) 2021 budget for the FS is $5.3 billion, including nearly $2.5 billion for wildland fire management and $2.8 billion for forest maintenance and management. In addition, the FS can access a Wildland Fire Management (WFM) reserve account or a Federal Land Assistance, Management, and Enhancement Act (FLAME) reserve account. The 2021 Federal budget includes $2.04 billion for these two accounts to increase wildfire suppression operations, if required.[8]

Before 1905, the DOI managed the Bureau of Forestry, which later moved to the USDA via the Federal Forest Transfer Act of 1905 and was renamed the Forest Service.[9] In recent years, there have been suggestions of moving the FS back into DOI to enhance wildfire management. The U.S. Government Accountability Office (GAO) released a report in 2009 indicating that such a move could align "federal land management missions . . . creating the opportunity for greater long-term program effectiveness." According to the report, the FS manages more than 190 million acres of land, and the agencies under DOI manage almost 450 million acres, all of which are either intermingled or adjacent.

Warnings about organizational issues and possible disruptions from such a move arose immediately. In the view of the FS, "a move could compromise the agency's work with state and private landowners, and with other USDA agencies, in carrying out USDA's mission to protect and enhance natural resources on private as well as public lands." A GAO report resurrected a potential reorganization plan discussed in 1979 that estimated $160 million ($575 million in 2022 dollars) of taxpayer savings over several years by moving the FS into DOI (see fig. 4).[10, 11]

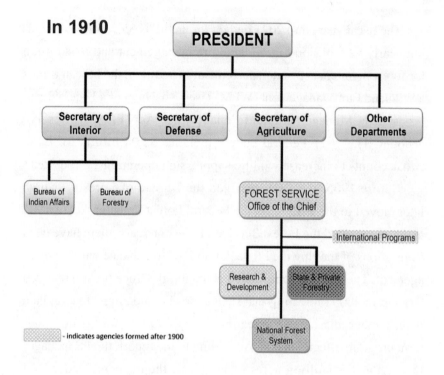

Figure 4. Organizational relationship of forest and wildfire management agencies in 1910. Congress authorized the Bureau of Forestry to be transferred to the United States Department of Agriculture in 1905. (Figure created by Crowbar Research Insights LLC™, © Crowbar Research Insights LLC™. All rights reserved.)

THE DEPARTMENT OF INTERIOR

Several bureaus within the DOI assist in wildland management for more than 400 million acres (see fig. 5). The first is the National Park Service (NPS), created in 1916, which oversees national parks, monuments, and other recreational areas that span more than 85 million acres. The bureau contains the Division of Fire and Aviation Management, which comprises the Aviation, Structural Fire, and WFM branches. The WFM branch is responsible for the fire management plans used by all parks with burnable vegetation. This branch also oversees the Wildland Fire Program. To achieve its mission, the WFM branch developed the

2020–2024 Wildland Fire Strategic Plan led by the Fire Management Leadership Board. The 2021 budget request for the NPS was $2.5 billion, including $4.0 million to reduce wildfire risk through active forest management mitigation work.[12-14] This branch also can access the WFM and FLAME reserve funds described earlier in the U.S. Department of Agriculture section.

Figure 5. The DOI building on the Mall in Washington, DC. (Photograph by Pgiam via iStock by Getty Images, March 4, 2013, https://www.istockphoto.com/photo/department-of-the-interior-gm174874795-23358468.)

Under the DOI is the Bureau of Indian Affairs (BIA), established in 1824 to manage and protect the United States' trust responsibility to American Indian and Alaska Native people, along with 55 million acres of land. This bureau also contains a Division of Forestry and Wildland Fire Management, which supervises the Branch of Forest Resource Planning and WFM. The Branch of Wildland Fire Management is responsible for wildfire response on Indian reservation managed land through the Wildland Fire Management Program. The proposed 2021 budget for the BIA included $54.1 million for the Tribal Forestry programs.[15,16]

Established in 1946, The Bureau of Land Management (BLM) is responsible for more than 245 million acres of public land. Within the BLM is the Fire and Aviation Program. The national BLM, along with state and field offices, administer this program with field offices overseeing on-the-ground fire management and aviation activities. The Fire and Aviation Program's headquarters are at the National Interagency Fire Center (see discussion below). The program partners with other federal and state agencies for wildfire management. The proposed 2021 BLM budget included $10.3 million for forest management on public lands.[17,18]

The U.S. Fish and Wildlife Service (FWS) manages 146 million acres of land in national wildlife refuges in all fifty states and the U.S. territories. Its fire management program administers wildfire management and protects more than 75 million burnable acres. The 2021 FWS budget request was $2.8 billion.[19,20]

Created in 2001 within the DOI, the Office of Wildland Fire oversees the Wildland Management Program. This program includes the four previously mentioned bureaus (NPS, BIA, BLM, and FWS) and 535 million public and tribal lands acres. This office is responsible for providing a unified program across the DOI. The 2021 Office of Wildland Fire budget request was $1.0 billion, including wildfire preparedness and suppression funding, an increase of $48 million from FY 2020.[21-23]

Between 1910 and 1950, the federal bureaucracies inside the DOI grew. During this same time, state legislatures established their forest management departments to protect state-owned lands (see fig. 6).[24,25]

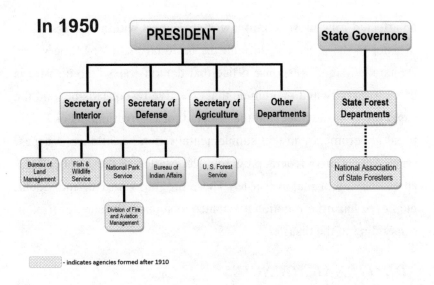

Figure 6. Organizational relationship of forest and wildfire management agencies in 1950. Congress authorized three new agencies and one new division between 1910 and 1950. States founded their forest departments starting in 1913. (Figure created by Crowbar Research Insights LLC™, © Crowbar Research Insights LLC™. All rights reserved.)

DEPARTMENT OF HOMELAND SECURITY

The Federal Emergency Management Agency (FEMA) is housed within the Department of Homeland Security, and its mission is "to help people before, during and after disasters."[26] Within FEMA is the U.S. Fire Administration (USFA), whose purpose is to help the public prepare for wildfires and assist those impacted by wildfires. The USFA supports fire and emergency medical services and oversees the National Fire Academy. FEMA's FY 2021 requested budget was $14.5 billion.[27]

FEMA provides federal assistance through the 1988 Robert T. Stafford Disaster Relief and Emergency Assistance Act. Like other bureaucratic entities, those requesting help must navigate an intricate process. First, local and state governments must determine that their resources are inadequate to meet the needs, save lives, and protect property. Then the state's governor or the tribal chief executive of an affected tribe must submit a written request within 30 days of the disaster to the president through the regional FEMA or Emergency Preparedness and Response office that demonstrates "the disaster is of such severity and magnitude that effective response is beyond the capabilities of the State and the affected local governments or Indian tribal government and that supplemental federal assistance is necessary." Activation of federal programs occurs after the president declares an emergency or major disaster. These programs provide individual, public, or hazard mitigation assistance to support the state or tribe in responding to the disaster.[28, 29]

ADDITIONAL AGENCIES

In 1965 the FS, BLM, and National Weather Service created the Boise Interagency Fire Center (BIFC) for working together "to reduce the duplication of services, cut costs, and coordinate national fire planning and operations." A few years later, the NPS, BIA, and FWS joined BIFC, and the name changed to the National Interagency Fire Center (NIFC). The primary focus of NIFC is managing wildland fires on more than 676 million acres and assisting in other emergencies. While the NIFC does not make policy decisions, the NIFC partner agencies work together on policy and training and share equipment, supplies, and personnel needed to fight wildfires (see fig. 7).[30]

Figure 7. The National Interagency Fire Center campus in Boise, Idaho. (Photo Courtesy of Leo Anthony Geis / Idaho Airships, Inc.)

Within the NIFC is the National Interagency Coordination Center (NICC), which depends on local resources to fight wildland fires. Some fires become too large for local management and require assistance from the nearest Geographical Area Coordination Center (GACC). There are ten GACCs throughout the United States and Alaska, comprised of regional federal and state fire directors from within each area. The GACCs provide local agency shared support and resources, including people, ground equipment, and aircraft. After exhausting these resources, the NICC steps in to coordinate the wildland fire response and allocates state, federal, and private resources. In addition, NICC is the dispatch center for most of the larger firefighting equipment and specialty-trained response teams. The National Multi-Agency Coordinating Group (NMAC) within NICC allocates the U.S., international, and military resources. NMAC includes representatives from the NIFC partner agencies, FEMA, and the National Association of State Foresters (NASF). Its mission is "National wildland fire operations management, priority setting, and resource allocation through multi-agency coordination.[31, 32]

The secretaries of agriculture and the interior established the Wildland Fire Leadership Council (WFLC) in 2002. This interagency committee consistently implements the National Fire Plan's goals, actions, and policies, along with the Federal Wildland Fire Management Policy, through the National Strategic Committee (NSC). The NSC executes WFLC's activities, including the National Wildland Fire Cohesive Strategy.[33-35]

Established in 1976 through a memorandum of understanding between the Departments of Agriculture and Interior, the National Wildfire Coordinating Group (NWCG) provides national leadership for wildland fire management. Nine agencies are members of NWCG, including BIA, BLM, FWS, FS, International Association of Fire Chiefs, Intertribal Timber Council, NASF, NPS, and USFA.

NWCG's purpose is explicitly stated in a memorandum:

> To establish an operational group designed to coordinate programs of the participating agencies to avoid wasteful duplication and to provide a means of constructively working together. Its goal is to provide more effective execution of each agency's fire management program. The Group provides a formalized system to agree upon training standards, equipment, aircraft, suppression priorities, and other operational areas. Agreed upon policies, standards, and procedures are implemented directly through regular agency channels.

Even though its goal is decreasing federal duplication, NWCG represents another level of bureaucracy, as it shares similarities to other national groups, including NIFC, NICC, and NMAC.[36]

The DOI agencies (BIA, BLM, FWS, and NPS) and the USDA's FS continued the bureaucratic path by signing another wildland fire management interagency agreement in 2017. "Because wildland fire recognizes no boundaries, the agencies must continually strive to provide interagency cooperation to achieve more productive, cost-effective and efficient operations among the partnering agencies," the agreement stated.

These agencies are responsible for developing and conducting wildland fire management plans through cooperatively sharing personnel, equipment, and supplies (see fig. 8).[37]

THE ROLE OF FEDERAL AND STATE AGENCIES

> *Authors' Note*
>
> The following diagram and organization chart attempts to sort out this convoluted maze. As we developed this chart, we could not help but summon our anger and disbelief that our elected representatives in Congress allowed agency bureaucrats to establish an organization no successful private or public company could or would tolerate.
>
> Little wonder why such a dysfunctional structure creates a lack of accountability, accessibility, and timely action.

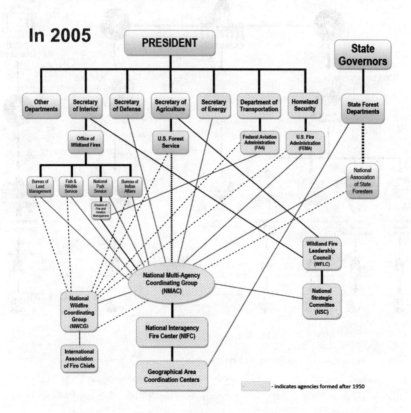

Figure 8. Organizational relationship of forest and wildfire management agencies in place in 2005 and continuing through today. In the 55 years since 1950, Congress authorized these new departments and divisions, agencies, councils, and programs. (Figure created by Crowbar Research Insights LLC™, © Crowbar Research Insights LLC™. All rights reserved.)

FUNDING GOVERNMENT OPERATIONS

The following schematic summarizes taxpayer funding of federal and state agencies described in figure 8. Taxpayer-elected representatives are responsible for funding and overseeing each of the agencies and bureaucracies they create (see fig. 9).

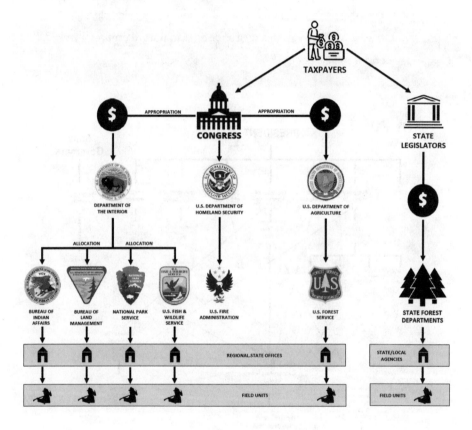

Figure 9. Federal and State Forest and Wildfire Management Organization Structure, 2020. (Chart adapted by Crowbar Research Insights LLC™ from a similar chart on the Office of Wildland Fires website, www.doi.gov/wildlandfire/budget, Crowbar Research Insights LLC™, © Crowbar Research Insights LLC™. All rights reserved.)

TAXPAYER COMMITMENT TO FOREST MAINTENANCE AND WILDFIRE MANAGEMENT

The following table contains 2020–2021 budget information, total acreage each agency is responsible for managing, and budgeted costs, where available, to preserve our national forests and suppress wildfires (see table 1 and fig. 10).

FY 2021/2022	Acres Managed (in Millions)	Total Annual Budget (in Billions)	Forest Management Annual Budget (in Billions)	Forest Management Percent Annual Budget	Forest Management Annual Cost Per Acre Managed
Interior	417	$ 27.1	$ 7.578	27.96%	$ 18.17
Agriculture	193	$ 198.0	$ 9.161	4.63%	$ 47.47
Homeland Security	-	$ 90.8	$ 0.053	0.06%	n/a
Defense/Energy/and All Others	30	$ 5,695.1	0	0.00%	$ -
Federal Agencies Totals	640	$ 6,011.0	$ 16.792	0.28%	$ 26.24
All State Agencies Totals	1,982	$ 2,113.8	$ 5.830	0.28%	$ 2.94
All Local Agencies Totals		$ 1,143.0			
All Agencies	2,622	$ 9,267.8	$ 22.622	0.24%	$ 8.63

Table 1. Federal, state, and local agencies wildfire and forest management FY 2021–2022 budgets. (Table created by Crowbar Research Insights LLC™ with data from Fiscal year 2022 budget justifications for federal agencies, latest state budgets for 2020–2021, acres managed from Congressional Research Service, https://sgp.fas.org/crs/misc/R42346.pdf, USDA Forest Service State and Private Forestry Fact Sheets, https://www.fs.usda.gov/, © Crowbar Research Insights LLC™. All rights reserved.)

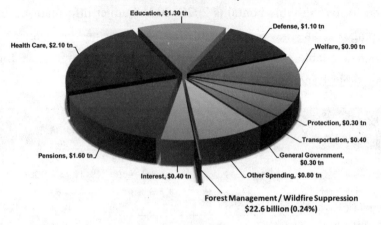

Figure 10. 2021–2022 total federal, state, and local agencies spend. (Figure created by Crowbar Research Insights LLC™ with information extracted from http://usgovernmentspending.com, © Crowbar Research Insights LLC™. All rights reserved.)

Based on our analysis of the 2021–2022 budget, the DOI, USDA, Homeland Security, Defense/Energy, and others will spend approximately $16.8 billion on forest maintenance and wildfire management. This figure represents only 0.28 percent of the total federal budget. Despite $8.5 billion of increased allocations since 2000, the number of burned acres of forestland also increased by more than 75 percent during the same period. Even with the most recent ten-year funding from the 2021 Infrastructure Investment and Jobs Act, the federal funding commitment is not keeping pace. It is, quite frankly, an embarrassment, considering the stated priorities of preserving our forests. Americans impacted by wildfires each year (212 million, or nearly 65 percent of the country's population) deserve better.[38]

STATE AND LOCAL AGENCIES

We also compiled each state's budget for forest management, compared it to the total state budget, and determined the cost per acre of land managed by the state. In addition, we compiled the costs in the fifteen most fire-prone states and compared these metrics to the remaining states.

With the exception of California, which increased its forest management budget significantly in 2021 and again in 2022, each of the fifteen most fire-prone states spends, on average, approximately 86 percent less than the federal government spends on a per-acre basis for forest management. Remember: forest management budgets include both wildfire suppression costs and the stewardship costs required to sustain our forests.

Chapter 4 examines these fifteen fire-prone states in more detail—and with good reason. Increased focus on these states' funding mechanisms, equipment, personnel requirements, and viable solutions can produce long-term benefits. We developed many of our recommendations to address the problems in these fifteen states directly from our analyses (see tables 2 and 3).

Fifteen Most Fire-Prone States	State Acres Managed (in Millions)	State Annual Budget (in Billions)	Forest Management Annual Budget (in Billions)	Forest Management Percent of State Annual Budget	Forest Management Annual Cost Per Acre Managed
1 Alaska	218.54	$ 9.50	$ 0.07	0.70%	$ 0.31
2 Arizona	40.48	$ 35.20	$ 0.02	0.05%	$ 0.44
3 California	56.60	$ 331.10	$ 4.43	1.34%	$ 78.27
4 Colorado	88.59	$ 30.40	$ 0.04	0.12%	$ 0.41
5 Florida	38.19	$ 93.40	$ 0.10	0.11%	$ 2.74
6 Georgia	53.16	$ 43.80	$ 0.06	0.13%	$ 1.04
7 Idaho	25.94	$ 9.80	$ 0.01	0.05%	$ 0.20
8 Montana	30.14	$ 7.30	$ 0.01	0.20%	$ 0.48
9 Nevada	40.99	$ 14.60	$ 0.05	0.37%	$ 1.31
10 Oklahoma	18.11	$ 21.50	$ 0.02	0.07%	$ 0.87
11 Oregon	45.65	$ 39.70	$ 0.15	0.37%	$ 3.23
12 Texas	219.75	$ 146.20	$ 0.06	0.04%	$ 0.27
13 Utah	33.47	$ 20.30	$ 0.03	0.13%	$ 0.80
14 Washington	34.44	$ 46.70	$ 0.15	0.31%	$ 4.24
15 Wyoming	39.66	$ 4.80	$ 0.01	0.22%	$ 0.26
Sub total	983.69	$ 854.30	$ 5.19	0.61%	$ 5.28

All Other States	State Acres Managed (in Millions)	State Annual Budget (in Billions)	Forest Management Annual Budget (in Billions)	Forest Management Percent of State Annual Budget	Forest Management Annual Cost Per Acre Managed
16 Alabama	51.50	$ 29.90	$ 0.02	0.07%	$ 0.42
17 Arkansas	37.35	$ 20.20	$ 0.03	0.14%	$ 0.76
18 Connecticut	4.15	$ 24.30	$ 0.00	0.01%	$ 0.39
19 Delaware	0.91	$ 9.80	$ 0.00	0.02%	$ 1.79
20 Hawaii	3.77	$ 15.00	$ 0.05	0.36%	$ 14.28
21 Illinois	15.52	$ 69.70	$ 0.01	0.01%	$ 0.39
22 Indiana	12.10	$ 36.10	$ 0.01	0.04%	$ 1.05
23 Iowa	10.04	$ 20.90	$ 0.00	0.01%	$ 0.30
24 Kansas	48.86	$ 17.00	$ 0.00	0.02%	$ 0.08
25 Kentucky	29.48	$ 31.10	$ 0.02	0.05%	$ 0.57
26 Louisiana	33.93	$ 32.20	$ 0.01	0.05%	$ 0.44
27 Maine	19.29	$ 9.10	$ 0.01	0.11%	$ 0.54
28 Maryland	5.65	$ 43.80	$ 0.01	0.02%	$ 1.68
29 Massachusetts	6.18	$ 63.80	$ 0.00	0.01%	$ 0.61
30 Michigan	40.17	$ 63.80	$ 0.03	0.05%	$ 0.80
31 Minnesota	62.71	$ 37.90	$ 0.04	0.10%	$ 0.62
32 Mississippi	38.40	$ 18.10	$ 0.03	0.16%	$ 0.76
33 Missouri	57.64	$ 31.60	$ 0.02	0.05%	$ 0.28
34 Nebraska	49.52	$ 9.10	$ 0.01	0.06%	$ 0.11
35 New Hampshire	9.35	$ 7.70	$ 0.00	0.03%	$ 0.28
36 New Jersey	5.24	$ 70.50	$ 0.01	0.01%	$ 1.21
37 New Mexico	64.00	$ 17.90	$ 0.01	0.07%	$ 0.20
38 New York	41.76	$ 168.30	$ 0.02	0.01%	$ 0.43
39 North Carolina	36.25	$ 54.10	$ 0.07	0.12%	$ 1.82
40 North Dakota	36.55	$ 5.90	$ 0.01	0.13%	$ 0.21
41 Ohio	13.63	$ 74.40	$ 0.01	0.02%	$ 0.84
42 Pennsylvania	33.62	$ 96.30	$ 0.03	0.03%	$ 0.98
43 Rhode Island	0.79	$ 9.00	$ 0.00	0.02%	$ 2.20
44 South Carolina	26.51	$ 30.00	$ 0.04	0.13%	$ 1.51
45 South Dakota	50.82	$ 4.80	$ 0.01	0.14%	$ 0.13
46 Tennessee	39.62	$ 29.60	$ 0.03	0.11%	$ 0.80
47 Vermont	9.22	$ 5.30	$ 0.00	0.07%	$ 0.41
48 Virginia	31.48	$ 52.50	$ 0.03	0.07%	$ 1.10
49 West Virginia	24.02	$ 13.90	$ 0.01	0.06%	$ 0.33
50 Wisconsin	48.26	$ 35.90	$ 0.03	0.08%	$ 0.60
Other State Agencies	998.32	$1,259.50	$ 0.62	0.05%	$ 0.62
State Agencies Totals	1,982.01	$2,113.80	$ 5.83	0.28%	$ 2.94

Table 2. State acres managed and 2021-2022 budgets. (Table created by Crowbar Research Insights LLC™ with data from USDA Forest Service State and Private Forestry Fact Sheets, https://www.fs.usda.gov/, © Crowbar Research Insights LLC™. All rights reserved.)

THE ROLE OF FEDERAL AND STATE AGENCIES

Fifteen Most Fire-Prone States	State Acres Managed (in Millions)	State Annual Budget (in Billions)	Forest Management Annual Budget (in Billions)	Forest Management Percent of State Annual Budget	Forest Management Annual Cost Per Acre Managed
1 Alaska	218.54	$ 5.60	$ 0.04	0.70%	$ 0.18
2 Arizona	40.48	$ 10.40	$ 0.01	0.06%	$ 0.14
3 California	56.60	$ 84.40	$ 0.22	0.27%	$ 3.96
4 Colorado	88.59	$ 10.20	$ 0.01	0.12%	$ 0.14
5 Florida	38.19	$ 31.10	$ 0.04	0.12%	$ 0.98
6 Georgia	53.16	$ 17.60	$ 0.02	0.14%	$ 0.45
7 Idaho	25.94	$ 3.20	$ 0.00	0.06%	$ 0.07
8 Montana	30.14	$ 3.00	$ 0.01	0.21%	$ 0.21
9 Nevada	40.99	$ 3.80	$ 0.02	0.40%	$ 0.37
10 Oklahoma	18.11	$ 7.50	$ 0.01	0.08%	$ 0.31
11 Oregon	45.65	$ 11.90	$ 0.05	0.40%	$ 1.05
12 Texas	219.75	$ 44.20	$ 0.02	0.05%	$ 0.10
13 Utah	33.47	$ 6.60	$ 0.01	0.14%	$ 0.28
14 Washington	34.44	$ 19.50	$ 0.06	0.32%	$ 1.83
15 Wyoming	39.66	$ 1.70	$ 0.00	0.22%	$ 0.09
Sub total	**983.69**	**260.70**	**0.52**	**0.20%**	**$ 0.53**

All Other States	State Acres Managed (in Millions)	State Annual Budget (in Billions)	Forest Management Annual Budget (in Billions)	Forest Management Percent of State Annual Budget	Forest Management Annual Cost Per Acre Managed
16 Alabama	51.50	$ 12.00	$ 0.01	0.08%	$ 0.19
17 Arkansas	37.35	$ 6.90	$ 0.01	0.15%	$ 0.28
18 Connecticut	4.15	$ 13.40	$ 0.00	0.01%	$ 0.23
19 Delaware	0.91	$ 3.40	$ 0.00	0.02%	$ 0.67
20 Hawaii	3.77	$ 6.40	$ 0.02	0.38%	$ 6.44
21 Illinois	15.52	$ 29.10	$ 0.00	0.01%	$ 0.17
22 Indiana	12.10	$ 13.60	$ 0.01	0.04%	$ 0.43
23 Iowa	10.04	$ 8.20	$ 0.00	0.02%	$ 0.13
24 Kansas	48.86	$ 6.30	$ 0.00	0.02%	$ 0.03
25 Kentucky	29.48	$ 12.40	$ 0.01	0.06%	$ 0.24
26 Louisiana	33.93	$ 12.80	$ 0.01	0.05%	$ 0.19
27 Maine	19.29	$ 4.50	$ 0.01	0.12%	$ 0.29
28 Maryland	5.65	$ 15.00	$ 0.00	0.02%	$ 0.62
29 Massachusetts	6.18	$ 23.20	$ 0.00	0.01%	$ 0.24
30 Michigan	40.17	$ 25.50	$ 0.01	0.05%	$ 0.34
31 Minnesota	62.71	$ 15.70	$ 0.02	0.11%	$ 0.27
32 Mississippi	38.40	$ 7.70	$ 0.01	0.17%	$ 0.34
33 Missouri	57.64	$ 12.80	$ 0.01	0.06%	$ 0.12
34 Nebraska	49.52	$ 4.20	$ 0.00	0.06%	$ 0.05
35 New Hampshire	9.35	$ 3.30	$ 0.00	0.04%	$ 0.13
36 New Jersey	5.24	$ 26.10	$ 0.00	0.01%	$ 0.48
37 New Mexico	64.00	$ 6.30	$ 0.00	0.07%	$ 0.07
38 New York	41.76	$ 65.70	$ 0.01	0.01%	$ 0.18
39 North Carolina	36.25	$ 20.30	$ 0.03	0.13%	$ 0.74
40 North Dakota	36.55	$ 2.30	$ 0.00	0.14%	$ 0.09
41 Ohio	13.63	$ 31.70	$ 0.01	0.02%	$ 0.37
42 Pennsylvania	33.62	$ 36.30	$ 0.01	0.04%	$ 0.40
43 Rhode Island	0.79	$ 4.00	$ 0.00	0.02%	$ 1.04
44 South Carolina	26.51	$ 12.40	$ 0.02	0.14%	$ 0.65
45 South Dakota	50.82	$ 2.00	$ 0.00	0.15%	$ 0.06
46 Tennessee	39.62	$ 12.50	$ 0.01	0.11%	$ 0.35
47 Vermont	9.22	$ 2.30	$ 0.00	0.07%	$ 0.18
48 Virginia	31.48	$ 17.20	$ 0.01	0.07%	$ 0.39
49 West Virginia	24.02	$ 6.20	$ 0.00	0.06%	$ 0.16
50 Wisconsin	48.26	$ 14.70	$ 0.01	0.09%	$ 0.27
Other State Agencies	**998.32**	**$ 496.40**	**0.26**	**0.05%**	**$ 0.26**
State Agencies Totals	**1,982.01**	**$ 757.10**	**0.782**	**0.10%**	**$ 0.39**

Table 3. State acres managed and 2001-2000 budgets. (Table created by Crowbar Research Insights LLC™ with data from USDA Forest Service State and Private Forestry Fact Sheets, https://www.fs.usda.gov/, © Crowbar Research Insights LLC™. All rights reserved.)

There are 2.27 billion acres of land in the United States. The federal government owns nearly 28 percent—approximately 640 million acres. State and local landmass cover about 193 million acres (8.5 percent), whereas private lands (owned by individuals, businesses, and trusts) amount to approximately 1.4 billion acres (61 percent).[39, 40] These acres are separate from any federal jurisdiction, including national forest management or fire-suppression responsibilities.

Each state began creating forestry departments and natural resources in the early twentieth century. Today all fifty states have taxpayer-funded departments managing the prevention and suppression of wildfires on lands falling under their jurisdiction, along with forest management and maintenance.

We believe tables 2 and 3 represent the first analysis of total spending on forests at the state level. They also summarize data from the fifteen most fire-prone states based on NICC reports. As outlined in these tables, state spending on forests during the fiscal year 2021–2022 amounted to a disappointing $2.94 per acre—larger than $0.39 per acre a decade previously, but still grossly insufficient. California is the only state to materially increase spending to $78.27 per acre in 2021–2022 from $3.96 per acre in 2000–2001. The other fourteen most fire-prone states only increased their combined spending to $0.82 per acre in 2021–2022 from $0.32 per acre in 2000–2001.

The University of Idaho compiled two surveys focused on the ten western states' wildfire suppression expenditures. According to its latest report, containing eleven years of data, wildfire suppression costs totaled $11.9 billion of the total forestry budget, or about $1 billion per year for all ten states from 2005 to 2015.[41] Based on the University of Idaho data compared with the same ten states' data developed in table 2, each state averages around 40 percent of its forestry department budget on wildfire suppression.

Because wildfires know no distinction between federal, state, private, or local land, the NASF, formed in 1920, helps coordinate state and federal wildfire data and response. Managed by the directors of the forestry agencies in the states, this nonprofit organization oversees the states, U.S. territories, and District of Columbia. Although NASF is not officially part of any state or federal organization, it appears as a signatory to many memoranda of understandings with federal wildfire agencies. More than 1.5 billion acres of state and private forests are protected through the coordination efforts of NASF.[42]

Municipal and volunteer fire departments are responsible for wildfire management at the local level. Unfortunately, financial and budget information from those levels was not readily available for public inspection. When combining the available federal and state information, results indicate that the United States spends $8.63 per year per acre to maintain, preserve, and protect our national forests, including costs to suppress wildfires.

By comparison, the average American family spends $2,200 per year per acre to maintain, preserve, and protect their home lawns and gardens.[43, 44] There's no misunderstanding what these figures represent: government agencies spend less than 1 percent per acre of the amount spent by the average American homeowner. A society is judged by how it protects and preserves its natural resources. Government priorities must change.

TOO MANY HANDS IN THE POT

The opportunity to become stewards of our national forests began nearly 120 years ago. Yet rather than sensibly managing our forests, bureaucrats have formed a maze of departments, bureaus, agencies, and councils only to have wildfires become more extensive and more intense—one more example of a truth that could not be more accurate today than it was when expressed some years ago by noted economist Milton Friedman: "The government solution to a problem is usually as bad as the problem."[45]

Why have fiscal priorities strayed away from our country's founding principles? Each fire season diminishes our collective ability to enjoy our treasured forests. The status quo of annual uncontrolled wildfire events caused by government ineptitude can never be an acceptable option.

THE KEY TAKEAWAYS FROM THIS CHAPTER ARE:

- Too many agencies, bureaus, programs, and groups are charged with managing our forests to the extent that understanding which organization is responsible for which piece of the forest management puzzle is overwhelming and confusing.

- Congress must act to force simplification of organizations and their responsibilities.

- Federal and state budgets are woefully inadequate for today's forest and wildfire management requirements.

- The fifteen most fire-prone states, with the exception of California, spend on average 86 percent less than the already depressing minimal spending by the federal government on state forest stewardship.

- The average American family spends more to maintain their lawns and gardens than federal and state governments spend protecting our treasured forests.

- It is time to challenge government—the status quo is no longer an option!

CHAPTER 3

COST OF WILDFIRES AND FORESTLAND MANAGEMENT

A billion acres of land across America are at risk of catastrophic wildfires . . . which are now a year-round phenomenon.

—*Vicki Christiansen, immediate past chief, USDA Forest Service*[1]

As wildfires decimate our national forests and become year-round events, costs will invariably burgeon. As outlined in the previous chapter, only a little imagination is needed to project a continuum of 40 percent increases in federal and state spending every twenty years to fight wildfires while managing our forests. This escalation is grossly underestimated.

This chapter uses the government agencies' budgets detailed in chapter 2 as one element to determine the annual cost of wildfires and forest management. Government agency costs are only a portion of the yearly expenses of managing wildfires and forests. As described in later chapters, our forests require additional resources for long-term maintenance. Without these resources, the taxpayers' costs described in this chapter will dramatically escalate.

DIRECT COSTS

The extensive literature on forest management distinguishes between the costs associated with wildfire suppression, extinguishing the fire, and forest maintenance. Suppression costs include forest preparedness (timber management within forests and around wildland-urban interface [WUI] areas, undergrowth removal, prescribed burns) and the expenses incurred to extinguish wildfires. Maintenance costs consist of forest protection and restoration, thinning, soil erosion, and flooding. Other direct costs include property, infrastructure, evacuation, and disaster relief.

FEDERAL AND STATE AGENCY DIRECT COSTS

The charts below detail the total spending for federal and state agencies on wildfire suppression and forest management in 2021–2022 and 2000–2001 (see figs. 1 and 2).

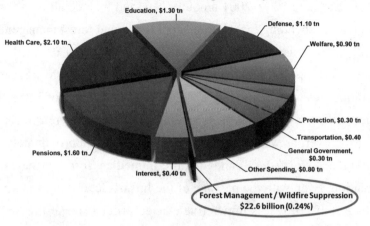

Figure 1. 2021–2022 total federal, state, and local agencies spend. (Figure created by Crowbar Research Insights LLC™ with information extracted from http://usgovernmentspending.com, © Crowbar Research Insights LLC™. All rights reserved.)

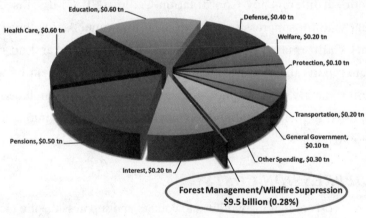

Figure 2. 2000–2001 total federal, state, and local agencies spend. (Figure created by Crowbar Research Insights LLC™ with information extracted from http://usgovernmentspending.com, © Crowbar Research Insights LLC™. All rights reserved.)

As clearly outlined in the above figures, the United States spends less than 1 percent of its annual resources on our forests. The

2021–2022 and 2000–2001 amounts reported include all local, state, and federal spending for wildfire suppression and forest management. While an increase to $22.6 billion in 2021–2022 from $9.5 billion in 2000–2001 appears significant, the stark reality remains. More than 10 million acres were destroyed in 2020 versus an average of 3.3 million acres in the forty previous years, ending in 2010 (see figure 2 in chapter 1). Moreover, as we will learn in a later chapter, approximately 100 million acres of federal lands require immediate forest management attention. Even with the passage of the Infrastructure Investment and Jobs Act of 2021, it is imperative elected officials make funding for wildfire suppression and forest management a higher priority.

Unfortunately, no local, state, or federal government agency separately accounts for wildfire suppression or forest-maintenance management costs. Unless we track these costs distinctly, we will not know the effectiveness of taxpayer spending. In early 2021, the National Interagency Coordination Center (NICC) disclosed annual wildfire suppression costs for the first time. As for the states, only California reports suppression costs annually, and no other state reports them as a component of forestry department budgets. Unfortunately, none of the agencies define the items included and reported in their annual budgets, making comparisons and analysis perplexing.

SUPPRESSION COSTS

Direct costs of fighting wildfires to contain and suppress the fire include personnel (salaries), equipment, supplies, and contracted services. Aircraft, helicopter, bulldozer, other heavy equipment, smoke jumpers, and firefighter costs are additional examples of suppression expenses.

Local, state, and federal agencies can and do react with human and equipment resources to fight wildfires that ignite on nonfederal (local, state, and private) lands. Since the federal government is responsible for responding to wildfires that flare up on federal lands, the distinction of who owns the land becomes essential because firefighting assets and personnel are initially deployed and paid for based on jurisdiction.

According to the Congressional Research Service, at the federal level, funding for wildfire suppression management has increased from $3.05 billion in 2012 to $6.11 billion in 2020 (amounts reported in inflation-adjusted FY 2020 dollars), an almost 100 percent increase (see fig. 3).[2]

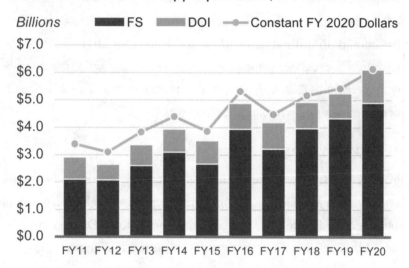

Figure 3. Federal government agency wildfire suppression costs. Data derived from annual Forest Service and Department of Interior appropriations acts, supplemental appropriations acts, committee reports, explanatory statements, and the detailed funding tables prepared by the House and Senate Committees on Appropriations. (Figure by Congressional Research Services, October 28, 2020, https://crsreports.congress.gov/product/pdf/R/R46583.)

In early 2021, the NICC published its first-ever summary of wildfire suppression costs. This action redefined wildfire suppression costs from historical precedents. Instead of reporting all suppression costs, only the actual costs to extinguish wildfires are reported. The revisions exclude expenditures related to forest preparedness, including timber management around WUI areas, undergrowth removal, and prescribed burns. Regardless of how wildfire suppression costs are defined, their surge is inevitable, as our forests face the grim reality of the future—more devasting wildfires.

For example, in 2001, wildfire suppression costs represented 37 percent of total federal forest management funding ($8.271 billion). In 2020 wildfire suppression costs increased to 52 percent of total federal funding for forest management and wildfire suppression. During the most recent five-year period (2016 to 2020), 306,604 wildfires on federal land burned 39 million acres, compared with 320,842 wildfires that burned 36 million acres of federal land from 2011 to 2015.[3] As these trends escalate, some project western states' wildfires to grow by 50 percent by 2050.[4,5] Simply put, the average size of wildfires is increasing—along with suppression costs.

PROPERTY LOSS

Property loss from wildfires includes damages to personal and business property, such as buildings, automobiles, recreational vehicles, business equipment and materials, and personal possessions. Damage may be caused by fire, smoke, or by firefighters during fire suppression—for example, water damage.[6] The chart below summarizes the average number of structures destroyed by decade compared to the last three years (see fig. 4). Note the 300 percent dramatic increase in the last column *just in the last three years*!

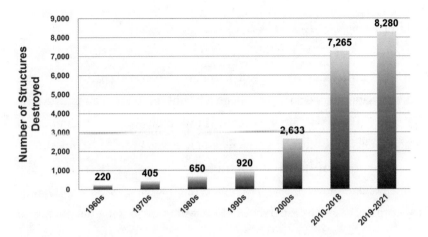

Figure 4. The average number of structures destroyed per year by wildfires. (Figure created by Crowbar Research Insights LLC™ with information extracted from the National Interagency Fire Center, © Crowbar Research Insights LLC™. All rights reserved.)

According to the National Fire Protection Association, direct property damage and losses from thirty-six fires in 2018 totaled more than $12.9 billion.[7] Three of the top-ten costliest U.S. wildland fires occurred in 2018. The Camp Fire, the most destructive in California history to date and the costliest fire in the United States, incurred $10.3 billion in insurance losses (FY 2020 dollars).[8] Insurance companies paid more than $24 billion in 2017 and 2018 combined from wildfire losses in California. This figure does not include the economic cost sustained by businesses for the wildfire interruption, time off work to repair damages, lost wages to the employees impacted, or damages caused by smoke and fine particulate matter, meaning the $24 billion amount is most likely a conservative estimate.[9]

Munich Re, a wildfire risk reinsurance firm, reported that property losses from wildfires in 2020 amounted to $16 billion.[10] For 2018, it reported wildfire property losses of $24 billion.[11] These claims are direct costs and do not include forest timber loss, increased insurance premiums, loss in property values, or other actual economic damages. A January 2021 report from the Swiss Re Institute corroborates these annual losses.[12]

UTILITY RATE INCREASES

Another direct cost resulting from wildfires is the rate increase passed on from utility companies to customers. Several California utilities recently announced a plan to spend $13 billion to reduce wildfire risk. Funding likely will be generated through customer rate increases. Pacific Gas and Electric (PG&E) already was granted permission to pass the costs of its wildfire mitigation efforts to its customers. A typical customer will see an average rate increase of $5.15 monthly. According to a spokesperson, PG&E wants to "reduce the chance of catastrophic wildfires."[13, 14] California is not the only state experiencing wildfire-related utility rate increases. Colorado's Xcel Energy requested a $58.6 million increase for power line maintenance to reduce wildfires caused by the lines.[15, 16] PG&E, which intends to bury 10,000 miles of power lines to mitigate wildfire risk, estimates the cost at $2.5 million per mile, or $25 billion for the total project.[17]

OTHER DIRECT COSTS

Infrastructure damage repair, evacuation aid, and disaster relief contribute to a wildfire's direct costs. Usually, disaster relief costs are incurred during and immediately after the wildfire.[18] They include temporary housing, lodging expenses, cleanup, repair and replacement costs, housing construction, childcare, medical expenses, household items, fuel, vehicles, moving expenses, and other comparable items.

Virtually all states reimburse local governments, impacted residents, and businesses for wildfire costs and damages, but the amounts are difficult to quantify. Unfortunately, multiple disaster relief accounts exist in state budgets, rendering analysis of actual out-of-pocket state expenditures impossible. Another obstacle is the possibility of double-counting disaster relief accounts in state and local budgets because transfers between these accounts are routine practice.

In June 2018 and May 2020, the Pew Charitable Trusts issued two separate studies focusing on how states pay for natural disasters. Pew cited the lack of transparency and accountability of taxpayer funds.[19,20] The two reports followed a 2015 U.S. Government Accountability Office (GAO) finding with similar conclusions.[21] We echo the findings of both GAO and Pew Charitable Trust studies and urge the negligent states identified in these reports to adopt sound accounting practices. Only through rigorous and consistent fiscal transparency can we assess the total direct costs of wildfires.

In its monthly reports, the Federal Emergency Management Agency (FEMA) tracks the fire disaster assistance declarations made to states for individual damage claims related to temporary housing, lodging, repair, medical, cleanup, and other direct costs caused by wildfires. Based on our review of the monthly reports from 2016 to 2020, FEMA provides about $2 million per year for wildfire damages.[22] While these reimbursements appear low, FEMA financing acts as "last-resort funding," coming into play only after exhausting all other sources of damage reimbursement, such as homeowner's insurance, medical coverages, and local and state disaster funding (see fig. 5).

Figure 5. The August Complex was the largest California wildfire, which began as thirty-seven separate wildfires within the Mendocino National Forest, set off after storms caused more than 10,000 lightning strikes. Approximately 10,500 structures were damaged or destroyed across California. (Photograph by Noah Berger/AP Photo, August 18, 2020, http://www.apimages.com/metadata/Index/APTOPIX-California-Wildfires/3021305b20e6 46e98bb6dfd77bf8629d/99/0.)

STATE AGENCIES WILDFIRE SUPPRESSION COSTS

Virtually every state maintains its forestry or natural resources department to manage forests and fight wildfires on state land. Unfortunately, these budgets are not always transparent, and many do not break down suppression and forest management costs separately. Moreover, the maze of state financial reporting practices described in the GAO and Pew Charitable Trusts reports make a detailed analysis of these practices impossible.

The University of Idaho compiled two surveys focused on ten western states' wildfire suppression expenditures. According to the university's report, consisting of eleven years of data, wildfire suppression costs totaled $11.9 billion of the total forestry budget, or about $1 billion per year, for all ten states from 2005 to 2015.[23] Using the University of Idaho data and comparing it to the same ten states' budget data we developed in chapter 2, each state spends about 40 percent of its forestry department budget on wildfire suppression, or approximately $2.5 billion annually, for all fifty states combined.

LOCAL GOVERNMENT AGENCY COSTS

Local governments are responsible for maintaining first-line firefighting and emergency medical response. Approximately 1.1 million professional men and women serve the country as first responders to any disaster or emergency impacting public health and safety.[24] When a wildfire breaks out in a WUI area, local first responders receive the initial 911 call.

The Census Bureau estimates local governments will spend approximately $53 billion for fire protection in FY 2021.[25] Local fire suppression, prevention, and maintenance costs are not reported separately as subcategories of this nationwide estimate; however, the National Interagency Fire Center reported that county, state, and private wildfires, which are under the responsibility of local and state government agencies, accounted for approximately 76 percent of all wildfires reported in 2020.[26]

Unfortunately, we could not assess the true annual monetary impact of local government wildfire and forest maintenance costs, because most states reimburse local government agencies for the incremental wildfire costs they incur. In 2019 California's Orange County received $7.2 million from the state for incremental wildfire firefighting costs incurred that year.[27] The reimbursement is reported as a cost in the California state budget.

There is an important nuance to this reimbursement practice, which our analysis of state agency costs described earlier in this chapter includes. States reimburse the local city and county firefighting agencies for out-of-pocket costs to fight wildfires on state lands. The opportunity exists for additional firefighter compensation—and corresponding base pay adjustments for future pension plan obligation determinations—for each wildfire subject to state reimbursement each year. If the state "foots the bill," costs can be reported under the shield of natural disaster relief.

INDIRECT COSTS

The total cost of fighting wildfires includes more than expenses from the actual direct suppression firefighting costs, referred to as indirect costs. Typically, other indirect and postfire costs and losses are more challenging to measure monetarily and may continue to accrue after extinguishing the wildfire. They include:

- loss of human life;

- private landowner fire prevention, preparedness, and education costs;

- economic loss to property values, business, and employment, particularly in fire-prone regions;

- recreational, tourism, and tax revenue loss;

- environmental loss through poor air and water quality;

- soil erosion and flooding;

- timber loss, environmental cleanup, and habitat restoration;

- health care for injuries sustained by firefighters and civilians;

- increased insurance premiums;

- loss of wildlife; and

- landscape mitigation to reduce future damage from flooding, landslides, erosion, and other issues related to areas burned by wildfires.[28]

In May 2018, Headwaters Economics prepared a report on community wildfire costs. This report was based on five wildfire case studies and existing literature on this topic covering specific wildfires in Colorado, California, and Arizona between 2002 and 2016. As summarized in figure 6, the Headwaters Economics report states that only 9 percent of the total wildfire costs examined resulted from suppression. Of the other 91 percent of costs, approximately 35 percent were from short-term expenses—home and property loss, evacuation, and relief aid costs—while the remaining 65 percent of wildfire costs were due to long-term damages (see fig. 6).[29]

Figure 1: Wildfire impacts as a proportion of total wildfire costs.

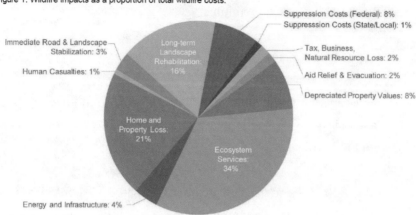

Figure 6. Wildfire impacts as a proportion of total wildfire costs. (Figure by Headwaters Economics, May 2018, https://headwaterseconomics.org/wp-content/uploads/full-wildfire-costs-report.pdf.) [Figure credit: Headwaters Economics 2018. Fully Community Costs of Wildfire. Used with permission from Headwaters Economics, headwaterseconomics.org.]

State or federal agencies, or a combination that can include local agencies, pay most suppression costs, as described earlier. "Nearly half of all wildfire costs are paid at the local community level by government agencies, non-governmental organizations (e.g., insurance), businesses, and homeowners. Almost all wildfire costs accrued at the local level result in long-term damages such as landscape rehabilitation, lost business and tax revenues, degraded ecosystem services, depreciated property values, and impacts to tourism and recreation," according to the report.[30]

Other studies arrive at similar conclusions. Each year, NOAA National Centers for Environmental Information (NCEI) prepares cost assessments for national disasters with direct and indirect economic losses exceeding $1 billion. Since 1980, there were eighteen wildfires with a total cost of $102.3 billion, meaning each event averaged $5.7 billion. Moreover, of the seven costliest disaster events, NCEI ranks wildfires as the third-highest cost on a per-event basis behind tropical cyclones and droughts.[31]

In one of the more comprehensive studies we examined, the National Institute of Standards and Technology (NIST) calculated the economic burden of wildfires on the U.S. economy. This calculation, consistent with the methodology we followed, included prevention and preparedness costs and direct, indirect, and postfire expenses and losses. Based on FY 2016 dollars, NIST estimated the economic burden of wildfires between $71.1 billion to $347.8 billion annually. Like the Headwaters Economics study, NIST estimated suppression costs at approximately 10 percent of total economic damages.[32]

Another repercussion of wildfires is the rising cost of fire insurance premiums that many homeowners are unable to afford. Additionally, some private insurance companies canceled their fire insurance policies. Home sales, particularly in WUI areas, may be adversely impacted by these costly premiums—or lack of available policies—as lenders will not finance mortgages without this insurance.[33]

Beyond the enormous annual economic and financial burden, wildfires in the United States also produce approximately 10 percent of the global wildfire greenhouse gas emissions (GHG) each year. Globally these emissions produce roughly twice as much CO_2 GHG emissions when compared with all the world's commercial aviation flights in 2019 and about 60 percent of the global emissions from automobiles.[34] Environmentalists argue that large wildfires will become the norm because of global warming.[35] We have not attempted to quantify the economic cost to the environment caused by wildfires. It obviously is quite substantial and one more reason to expeditiously extinguish them.

The following table summarizes our analysis of the annual cost of fighting wildfires and managing our forests (see table 1). We purposely excluded forest stewardship and management costs so that only wildfire damages are quantified.

Summary of Anual Wildfire Costs

Direct Costs:	Range ($Billion)
Federal Agency Suppression Cost	$6.0 to $7.0
Property Loss	$18.0 to $24.0
Utility Rate/Property Insurance Increases	$10.0 to $15.0
Stewardship/Forest Management	- -
State and Local Suppression Costs	$2.5 to $4.0
State and Local Stewardship Costs	- -
Subtotal Direct Costs	$36.5 $50.0
Indirect Costs:	
Infrastructure Damage, soil erosion/flooding, loss of life, property value loss, etc.	$50.0 to $25.0
Total Annual Cost of Wildfires	$86.5 $300.0

Table 1. Summary of annual wildfire costs. (Figure created by Crowbar Research Insights LLC™, © Crowbar Research Insights LLC™. All rights reserved.)

THE KEY TAKEAWAYS FROM THIS CHAPTER ARE:

- Each year, wildfires cost $86.5 billion to $300 billion in the United States, excluding the substantial adverse environmental impact from wildfire GHG emissions.

- Direct wildfire costs range from $36.5 billion to $50 billion each year.

- Indirect economic costs from wildfire losses amount from $50 billion to $250 billion annually.

- Wildfire suppression costs are only a small portion, approximately 10 percent, of the total costs associated with a wildfire.

- Multiple studies corroborate the annual direct costs and indirect economic impact of wildfires.

- Accounting and funding practices require immediate improvement to provide transparency of taxpayer funding related to suppression, forest management, and disaster relief costs at federal, state, and local government levels.

- Wildfires are the third costliest disaster each year behind tropical hurricanes and droughts and are the only ones we can control with proper management and additional resources.

- Wildfire GHG emissions are twice the amount emitted by all commercial airline flights and more than half of all automobile emissions each year.

- Congress must place funding of forest management and wildfire suppression among its highest priorities.

CHAPTER 4

WILDFIRE OCCURRENCES AND RESPONSES

When fires are not controlled through Initial Attack (usually within 24 hours) extended firefighting activities occur that generally involve the use of additional firefighting resources when such fires then grow large and complex.[1]

The numbers speak for themselves: Between 2012 and 2021, an average of 61,289 wildfires each year burned an average of 7.4 million acres (a little larger than the state of Massachusetts).[2] According to the USDA Forest Service (FS), wildfires burned almost 141 million acres of land managed by the FS (more than the landmass of California) between 1960 and 1999—a forty-year time span. It took only thirteen years to exceed this number. Between 2000 and 2013, nearly 161 million acres (about the size of Texas) were destroyed by wildfires.[3]

WILDFIRE STATISTICS

The National Interagency Fire Center (NIFC) collects information about wildland fires based on data from every state and territory from the National Interagency Coordination Center (NICC). The information includes a breakdown by the responsible federal agency, the fire's origin (human or lightning), number of acres burned, and the total number of fires and acres burned. Additionally, the NIFC compiles the total wildland fires and acres for each year between 1926 and 2020. In the last decade, 2011 saw the most wildland fires, with 74,126, while the highest total number of acres burned (more than 10 million) occurred in both 2017 and 2020.[4]

We developed the chart in figure 1 to summarize wildfire acres destroyed from 1960 to 2020. This chart includes the human- and lightning-caused wildfires over the 1992–2020 period. The NICC first reported lightning-caused data in 1992 (see fig. 1).

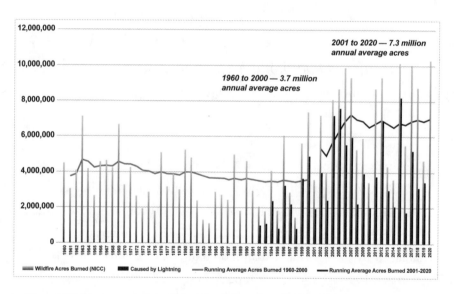

Figure 1. Wildfires consumed the United States acres from 1960 to 2020. Annual wildfire acres burned, 1960–2020, and lightning-caused acres burned, 1992–2020, with running average yearly. Lightning-caused fire reporting began in 1992. (Figure created by Crowbar Research Insights LLC™ with information extracted from https://www.nifc.gov/, © Crowbar Research Insights LLC™. All rights reserved.)

Based on our analysis of NICC-compiled data, lightning caused nearly 60 percent of all wildfires from 1992 to 2020. In media advertisements, "Smokey Bear," created by the FS, claimed that "humans cause nine out of every ten wildfires."[5] Instead, our analysis indicates humans cause four out of every ten wildfires nationwide. It is not our intent to discredit Smokey Bear. The critical point is weather forecasts can predict lightning-caused fires, and detection systems also can identify them. As discussed later in this chapter, weather forecasts and other predictive systems allow advanced firefighting assets to attack wildfire incidents immediately.

Wildfires impact over half the landmass of the United States (see fig. 2). The number of acres burned by wildfires has increased while the number of wildfire incidents has decreased slightly. Upon more careful examination, data revealed almost 60 percent of reported wildfire incidents in 2021 occurred east of the Mississippi River, impacting, on average, less than 29 acres per incident.[6] Since the early 2000s, this analysis holds true, as wildfire events rarely become disasters east of the Mississippi due to wetter weather. Data after 2013 also correlates with comparable metrics for the location of wildfire events. Significant wildfire events occur in western and southeastern states, but their impact is national and worldwide in scope.

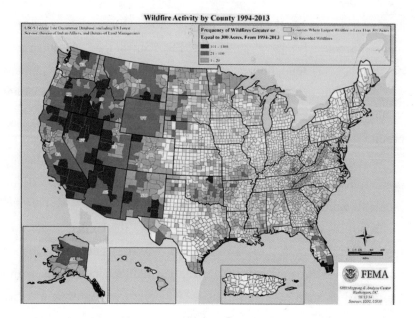

Figure 2. Wildfire activity by county, 1994–2013. (Figure by FEMA, https://www.dnr.wa.gov/publications/rp_fire_how_to_prepare_wildfire.pdf?uc1go.)

From 1960 through 2021, the top five years with the most nationwide acres burned by wildfires were 2006, 2007, 2015, 2017, and 2020. The second-highest number of acres burned (10.12 million), twice as many square miles as the combined square miles of the states of Rhode Island and Connecticut, with the smallest number of wildfires (59,000) occurred in 2020.[7] Wildfires are becoming more destructive (see fig. 3).

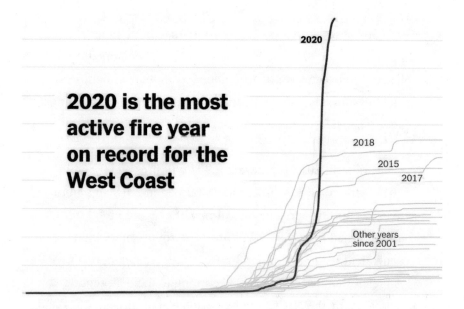

Figure 3. Record wildfires on the West Coast are capping a disastrous decade. (Figure by the *New York Times*, September 28, 2020, https://www.nytimes.com/interactive/2020/09/24/climate/fires-worst-year-california-oregon-washington.html.) [Photo credit: From the *New York Times*. © 2020 the New York Times Company. All rights reserved. Used under license.]

The locations where significant wildfire damages occurred in the fifteen most fire-prone states were examined. The highlights summarize the highest number of wildfires and the number of acres burned in each state in the past ten years.[8]

- Alaska: Wildland fires were highest in 2015 (768) followed by 720 fires in 2019. More than 5 million acres burned in 2015 and nearly 2.5 million in 2019. Alaskan wildfires emit carcinogenic particulate matter stretching across Canada, the United States, and Europe.

- Arizona: In 2016 Arizona reported 2,288 fires, and in 2018 the state reported 2,000 fires. An unusually high number of acres (slightly more than 1 million) burned in 2011, while

978,567 acres burned in 2020—more than three times the size of Phoenix. In 2013 the Yarnell fire tragically claimed the lives of nineteen Granite Mountain Hotshot crew members, and the subsequent investigation recommended implementing additional safety practices designed to protect firefighters.

- California: In 2013 there were 9,907 wildfires and 9,560 in 2017. More than 1.8 million acres burned in 2018 and more than 1.2 million acres in 2017. In 2020 a record of 4.3 million acres burned from 9,917 wildfires, destroying more than the equivalent landmass of San Francisco, Sacramento, and Los Angeles counties combined. According to incident reports published by the NICC, the Dixie and Caldor fires in 2021 and the August Complex Fire in 2020 each destroyed more than 1 million acres of forest. They also were California's most costly, with combined reported economic damages exceeding $1.5 billion, destroying over 1,850 structures. The Camp Fire in 2018 destroyed 85,000 structures and the entire town of Paradise, tragically claiming 85 lives. It stands as the costliest wildfire, with economic losses totaling $16.6 billion.

- Colorado: The highest number of wildfires (1,498) occurred in 2012, followed by 1,328 in 2018. The 2018 wildfires burned 475,803 acres, considerably more than the 246,445 acres in 2012. In 2020 625,000 acres (six times the size of Denver) were destroyed by 1,080 wildfires. The Marshall fire in January 2022 near Boulder destroyed nearly 1,000 structures, leaving an estimated economic loss of $1.6 billion.[9] The 2012 Waldo Canyon wildfire caused the evacuation of 32,000 Colorado Springs residents and the destruction of 346 homes.

- Florida: In 2011 4,736 wildfires consumed 253,746 acres. In 2017 2,380 wildfires burned 298,831 acres.

- Georgia: The Peach state had to combat 6,403 wildfires in 2011 and 5,086 in 2016. More than 200,000 acres burned in 2017, double the nearly 100,000 acres burned in 2011. Georgia had the fifth-highest number of wildfires in the United States that year.

- Idaho: Wildfires were highest in 2017 (1,598), followed by 1,180 in 2014. In 2012 1.6 million acres burned, an unusually high number, and 804,094 acres burned in 2015.

- Montana: The state experienced nearly the same number of wildfires in 2015 and 2017, with 2,432 and 2,422. More than 1.3 million acres burned in 2017, followed by 1.2 million in 2012. The Great Fire of 1910—commonly referred to as the Big Blowup, the Big Burn, or the Devil's Broom Fire—burned three million acres in north Idaho and western Montana, extending into eastern Washington and southeast British Columbia. The Great Fire is credited as the impetus for developing early wildfire prevention and suppression strategies.

- Nevada: In 2017 more than 1.3 million acres burned, and 1 million acres burned in 2018. There were 944 wildfires in 2012 and 807 in 2011.

- Oklahoma: In 2016 and 2017, the wildfire counts were similar, with 1,938 and 1,906. In 2016 767,780 acres burned, and 745,097 more in 2018.

- Oregon: The highest number of wildfires was recorded in 2014 (3,087), followed by 2,588 in 2015. More than 1.2 million acres burned in 2012 and an estimated one million in 2014. In 2020 Oregon suffered its second-highest destruction from wildfires as 1.142 million acres burned, equivalent to ten times the size of Portland. The 2021 Bootleg Fire burned nearly 415,000 acres and was the third-largest fire in the state's history since 1900 and the second-largest wildfire in the United States' 2021 wildfire season.

- Texas: Wildfires numbered 10,620 wildfires in 2012 and 10,541 in 2018. An unusually high number of acres burned in 2011 (more than 2.7 million), and 734,682 acres burned in 2017.

- Utah: In 2012 Utah reported 1,534 wildfires and 1,276 wildfires in 2013. The number of acres burned in 2018 and 2012 was similar: 438,983 and 415,266, respectively.

- Washington: Wildfires were highest in 2015, with 2,013, followed by 1,743 in 2018. More than 1.1 million acres burned in 2015, an unusually high number; 438,833 acres burned in 2018, and 842,370 acres—nine times the size of Seattle—burned in 2020.

- Wyoming: Approximately 838 wildfires were reported in 2012 and 711 in 2016. In 2012 357,117 acres burned; 279,242 acres burned in 2018; and 339,783 acres burned in 2020.

- Interestingly, the highest number of wildfires and acres burned in the same year occurred only four times in a ten-year period: Alaska in 2015, Oklahoma in 2016, Washington in 2015, and Wyoming in 2012.

> **How Are Wildfires Named?**
> Dispatchers and initial incident commanders designate a name to a fire. Typically, wildfires are named after the area where they start. A local landmark, geographical location, street, mountain, lake, or peak represent common sources. Notably, the name assigned provides an additional locator data point, allowing tracking by name and establishing priorities for fire suppression events.

The two charts that follow depict analyzed data to develop possible actions to extinguish wildfires faster in the fifteen most fire-prone states. Metrics include determining frequency and location of larger fires and whether these fires were sparked by lightning or were human-caused (see figs. 4 and 5). Our goal was to assess whether the historical data could provide insights into recommendations regarding wildfires' average size and location. Armed with this information, we believe we may be able to assess the most effective firefighting assets to be staged and deployed in these fifteen high-risk states.

To develop these charts, we accessed the FS spatial database for the annual time frames from 2010 to 2020. One chart displays information for human-caused wildfires and the other for lightning-caused wildfires. Each lists the fifteen most fire-prone states and categories for wildfire size: 300 to 1,000 acres, 1,001 to 10,000 acres, and more than 10,000 acres.

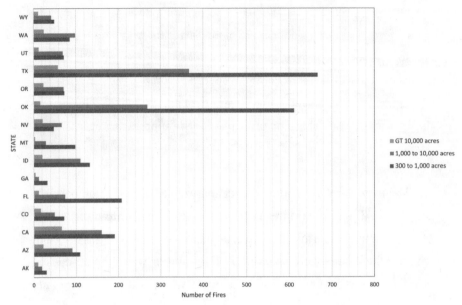

Figure 4. Human-caused wildfires in the fifteen most fire-prone states. GT means "greater than." (Figure created from research performed by Crowbar Research Insights LLC™, using Forest Service wildfire spatial databases, including https://www.fs.usda.gov/rds/archive/Catalog/RDS-2013-0009.5, https://web.archive.org/web/20201202170040/https://gacc.nifc.gov/sacc/predictive/intelligence/NationalLargeIncidentYTDReport.pdf, and https://www.fs.usda.gov/rds/archive/products/RDS-2013-0009.5/_fileindex_RDS-2013-0009.5.html, © Crowbar Research Insights LLC™. All rights reserved.)

The highlights from figure 4 include the following:

- First *and most important*: Human-caused wildfires in these states were *not* the most destructive in terms of large wildfires. For all fifteen states during the eleven-year period, 4,334 incidents occurred, consuming 20,422,000 acres (equivalent in size to the state of South Carolina) or, on average, about 1,850,000 acres per year for all fifteen states combined.

- The breakdown by size of wildfire is:

  » For 300 to 1,000 acres, there were 2,482 incidents involving 1.3 million acres, or an average of about 515 acres per incident.

  » For 1,001 to 10,000 acres, there were 1,528 incidents involving 4.5 million acres, or an average of about 3,000 acres per incident.

  » For 10,001 acres and above, there were 324 incidents involving 14.6 million acres, or an average of about 45,000 acres per incident.

- In the eleven-year period examined, Texas and Oklahoma had the highest number of human-caused wildfires, with 1,989 incidents, yet the average size of those particular wildfires was 240 acres each year per incident.

- California had even fewer human-caused wildfires at 417 total incidents for an average size of 910 acres per year per incident.

- The other twelve states combined had 1,928 incidents in the same time frame, with an average of 520 acres destroyed per year by human causes per incident.

While wildfire destruction causes personal and economic distress, from strictly a risk analysis point of view, annual human-caused fire events in these fifteen states appear to be managed appropriately without substantial new investment. It should be noted, however, that the FS classifies wildfires caused by utility power lines as "human-caused" events in their statistical summary reports. Events such as the 2018 Camp Fire in California caused by Pacific Gas and Electric power line sparks require separate action—in particular, the investment to bury power lines below ground.

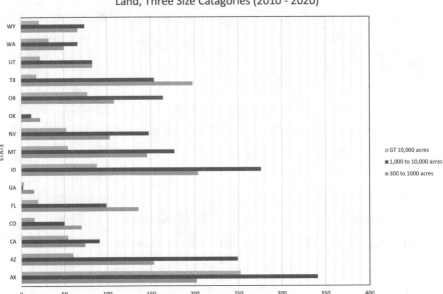

Figure 5. Lightning-caused wildfires in the fifteen most fire-prone states. GT means "greater than." (Figure created from research performed by Crowbar Research Insights LLC™, using Forest Service wildfire spatial databases, including https://www.fs.usda.gov/rds/archive/Catalog/RDS-2013-0009.5, https://web.archive.org/web/20201202170040/https://gacc.nifc.gov/sacc/predictive/intelligence/NationalLargeIncidentYTDReport.pdf, and https://www.fs.usda.gov/rds/archive/products/RDS-2013-0009.5/_fileindex_RDS-2013-0009.5.html, © Crowbar Research Insights LLC™. All rights reserved.)

The highlights from this figure include:

- Of particular importance, lightning-caused wildfires in these states were the most destructive. For all fifteen states during the time frame, 4,382 incidents occurred, consuming 39,800,000 acres (twice the acres for human-caused wildfires) or, on average, approximately 2.7 million acres per year.

- The breakdown by size of wildfire is:

  » For 300 to 1,000 acres, there were 1,630 incidents involving 900,000 acres, or an average of 550 acres per incident.

  » For 1,001 to 10,000 acres, there were 1,986 incidents involving 6.7 million acres, or an average of 3,400 acres per incident.

  » For 10,001 acres and above, there were 766 incidents involving 32.2 million acres, or an average of 42,000 acres per incident—more than twice the incidents and total acres compared to human-caused wildfires.

- During these eleven years, Alaska, Arizona, California, Idaho, Montana, Nevada, Oregon, and Texas had the highest number of lightning-caused wildfires, with 3,443 incidents and 34 million acres destroyed, approximately the size of the state of New York.

- Alaska leads all fifteen states with 795 total incidents and 12 million acres consumed in all three categories. Officials report an average event size of 10,001 acres and more events averaging 42,000 acres per incident.

- Georgia and Oklahoma had the fewest incidents, with 53 total reported events and a combined average size of 20,130 acres per incident.

In summary, lightning caused 65 percent of wildfire acre destruction in these fifteen most fire-prone states from 2010 to 2020. On the other hand, human and utility power line failures were responsible for 35 percent. Further, large lightning-caused wildfire events more than doubled the number of human and utility power line failure events.

Armed with the information from our analysis, the implications are as follows:

1. Current wildfire suppression practices are not effective.

2. Firefighting assets should be deployed where likely wildfire events occur—that is, the fifteen most fire-prone states.

3. When weather forecasts predict lightning, increased positioning of human and firefighting assets makes for prudent risk management.

4. In light of these findings, review and revision of past policy and operational practices are essential.

5. Extinguishing lightning-caused wildfires should become the top priority for all local, state, and federal agencies, given the vast positive impact such actions produce.

BEYOND THE NUMBERS: WILDFIRE RESPONSES AND RESOURCES

The response to a wildfire is dependent on whether the fire is burning on local, state, private land, tribal, federal, or a combination. Generally, local fire departments are the first responders after a 911 call reporting a wildfire. If the local resources are insufficient to extinguish the wildfire, state and federal assistance are secured from the nearest national Geographical Area Coordination Center (GACC).

The federal GACCs are responsible for the primary management coordination of forest maintenance and suppression on the 615 million acres of federal land under their care (see fig. 6). The agency also acts as the coordinating entity when wildfires cross jurisdictions or when state agencies require additional resources to suppress wildfires on lands under local or state responsibility (1.9 billion acres).

Figure 6. Location of the ten federal geographical area coordination centers. (Figure by the Geographical Area Coordination Centers, https://gacc.nifc.gov/.)

Here are some of the resources available to GACCs from federal agencies. The FS has several tools at its disposal to respond to wildfires. These include:

- Handcrews. Handcrews are the frontline firefighters, with twenty members in each crew. They construct "fire lines" to halt the advance of the fire.[10]

- Hotshots. These interagency crews manage the most challenging and sometimes the hottest areas of wildfires. They are sponsored by the FS, the National Park Service (NPS), Bureau of Land Management (BLM), Bureau of Indian Affairs (BIA), and state and county agencies.[11]

- Engine Crews. Fighting wildfires requires specialized equipment. These crews, consisting of two to ten firefighters, use firefighting equipment to spray water and foam. Engine crews also can serve as initial attack forces and assist with keeping wildfires from extending beyond established boundaries.[12]

- Smoke jumpers. These highly trained firefighters provide a quick initial attack on wildfires in remote areas. There are about 450 smoke jumpers in federal wildfire firefighting service. Using parachutes, they "jump" from airplanes into isolated locations, with their gear and supplies also dropped by parachute.[13]

- Mechanized equipment crews. These teams use bulldozers and other clearing equipment to establish fire perimeters, roads, and fire barriers.

- Helitack crews. Helicopters transport helitack crews to wildfires. The helicopters may land near the wildfire, or crews may rappel from hovering helicopters. They quickly assess and respond to wildfires by building fire lines.[14]

- Fire equipment. Firefighting equipment includes a variety of wildland fire engines specially designed for the type of wildfire and the affected terrain. Helicopters and airplanes transport firefighters, equipment, water, and fire retardant to wildfires.[15]

The FS, working with the NICC and GACCs, allocates resources based on an assessment of the complexity of each new fire using a rating scale from one to five (high to low), then activating additional firefighting assets to the closest dispatch area to the wildfire. A low-assessment fire may only need one fire engine, while a higher-ranked fire may require three engines. As the number of resources allocated to each fire grows depending on how quickly the wildfire spreads and the amount of life, property, and infrastructure affected, the assessment level increases. All local, state, and federal fire agencies responsible for the land impacted by the wildfire share these resources.[16] Complex fires with rankings of one to three require additional assets, including bulldozers, helicopters, and aerial firefighting aircraft.

In the BIA, wildland fire programs are managed by eighty-nine tribes and agency units either through the Indian Country's Wildland Fire Management Program or by self-determined or self-governance services. The BIA's fire operations section is responsible for managing the BIA's Wildland Fire Management Resource program.[17] Its wildland fire resources include:

- The BIA National Aviation Program. Aviation activities are supported through this program, which contracts aircraft and provides aviation crews and firefighters.[18]

- Helitack Program. Nine national helitack programs are available to protect tribal assets.[19]

- Engines. More than 200 engines and dozers, water, tenders, and heavy equipment are available to fight wildland fires. Three to seven firefighters comprise the engine crews.[20]

- Hotshot and Hand Crews. Each crew is comprised of 18-20 firefighters and is differentiated by experience levels. There are seven Interagency Hotshot Crews, who are the most experienced firefighters.[21]

Headquartered at NIFC, the BLM's Fire and Aviation Program works with the FS and six other federal agencies, including the BIA, to manage wildland fire. This program is responsible for protecting tribal and federal lands.[22] Aviation assets can be shared with states under formal prenegotiated memoranda of understanding to the extent they are available. The program's aviation resources include helicopters, single-engine air tankers, tactical air aircraft, utility aircraft, aerial supervision modules, heavy air tankers, smoke jumper aircraft, and large transport aircraft.[23] Federal and state agencies' aviation programs are discussed in more detail in chapter 6.

The NICC, working through the GACCs, also dispatches firefighting bulldozer equipment, supplies, crews, and aerial assets. The NICC is the sole federal dispatch center for "heavy air tankers, bulldozers, lead planes, smoke jumpers, hotshot crews, Type 1 Incident Management Teams, area command teams, medium and heavy helicopters, infrared aircraft, military resources, telecommunication equipment for fires, Remote Automated Weather Stations (RAWS), and large transport aircraft."[24]

NATIONAL INCIDENT MANAGEMENT SYSTEM

The Federal Emergency Management Agency created the National Incident Management System (NIMS) to guide "all levels of government, nongovernmental organizations, and the private sector to work together to prevent, protect against, mitigate, respond to, and recover from incidents." Adoption of the NIMS by organizations is required to be eligible for federal preparedness grants.[25]

The NIMS created the National Response Framework (NRF) guide to assist agencies in developing and integrating their response plans for disasters and emergencies. The NICC is a partner agency in the NRF and uses the NIMS to coordinate wildland fire emergency responses.[26,27]

There are three major components of the NIMS:

- resource management

- command and coordination

- communications and information management

The Incident Command System (ICS) falls within the Command and Coordination section.[28] The ICS manages people and resources required for fighting wildfires. Often, different agencies must work together, especially in the event of multijurisdictional wildfires. Called a Unified Command, this structure is designed to mitigate conflicting tactical instructions and direction when wildfires cross over onto state, federal, and tribal lands. Within the ICS are five wildfire assessment levels ranging from the least complex wildfire (Type 5) to the most complex wildfire (Type 1).[29] Incident command teams are assigned to individual wildfires based on training and experience. The most qualified teams are tasked with the most complex wildfires, while less senior teams are assigned to less complex incidents. Incident command teams for complex fires can reach fifty personnel, whereas Type 1-2 rated wildfires may have as few as five to ten team members.

The functional areas and leadership within the ICS include:

1. Incident command: Oversees the overall management of the incident. For every wildfire within a single jurisdiction or without agency overlap, an incident commander (IC) determines all strategies and tactics for managing the wildfire. If the wildfire is multijurisdictional, a unified command oversees the wildfire response. The unified command consists of ICs of the agencies involved to coordinate an effective response.

2. Operations: Conducts the operations to suppress the wildfire. The operations section chief manages the tactical operations of the strategy developed by the IC.

3. Planning: Collects and evaluates the information needed to develop the strategy and tactical operations.

4. Logistics: Coordinates all the incident support needs (services and supplies) required to assist the tactical operations excluding aviation support.

5. Finance and administration: Monitors all the financial aspects of the wildfire response. The section chief is responsible for providing financial information and advice to the IC. This function also collects interagency, state, tribal, and local firefighting cost data for later reimbursement, where applicable.

Three command staff assist the IC:

- The information officer delivers incident information to the appropriate agencies, incident personnel, and media.

- The liaison officer is the contact person with other involved agencies.

- The safety officer is responsible for developing and recommending measures to ensure the safety of all personnel involved in responding to the wildfire.[30-32]

THE CRITICALLY ESSENTIAL DISPATCHERS AND INCIDENT COMMANDERS

Dispatchers. Each 911 call involving a wildfire incident is routed to a dispatcher with wildland fire experience. According to the National Wildfire Coordinating Group (NWCG), each dispatcher's role includes the following key responsibilities, among others (see fig. 7).[33]

Prepare, Mobilize, and Build the Team
- Gather critical information pertinent to the assignment.
- Establish and communicate with the chain of command (particularly the incident commander), reporting procedures, risk management processes, and radio frequency management.
- Establish a common operating picture with the incident commander, supervisors, and subordinates.
- Identify assigned resources and maintain accountability.

Lead, Supervise, Direct
- Identify, analyze, and use relevant situational information to make informed decisions.
- Use, interpret topographic maps to plot locations.
- Retrieve and distribute appropriate intelligence such as spot weather and fire weather forecasts.

- Direct resources to plotted location.

Manage Risk
- Identify hazards.
- Make situational risk decisions.
- Implement controls.

Perform Initial Attack Dispatcher Duties, including:
- Work with local geographic area and national policies and guidelines.
- Use computer-aided dispatch systems following local protocols.
- Dispatch tactical resources according to hosting centers' procedures.
- Recognize and identify resource shortages.
- Follow up on confirmation of orders.
- Utilize contracts, agreements, and other existing sources to obtain resources.
- Monitor aircraft operations for adherence to regulations and safety procedures.
- Participate in regular briefings.

Figure 7. Dispatcher's responsibilities. (Summarized from the National Wildfire Coordinating Group Initial Attack Dispatcher Job Description, https://www.nwcg.gov/positions/iadp.)

Dispatchers receive appropriate on-the-job training and become qualified only after taking written and field observational examinations. They require only one year of fire management experience at the federal level. A dispatcher's role means responding quickly and accurately to each wildfire event with the appropriate firefighting assets while ensuring continuous communication with the IC also assigned to the wildfire.

Incident commanders. These individuals are responsible for structuring, organizing, and managing each wildfire event. They assess risk and benefit by applying sound firefighting and rescue practices. In short, they oversee and are the ultimate command and control decision-maker for each wildfire from its beginning until it is extinguished. Depending on the size and complexity of the wildfire, the incident commander assembles and supervises additional personnel to assist in carrying out critical functions. According to the NWCG, the key responsibilities of the IC include (see fig. 8):

Preseason Preparation
- Recruit and roster team before assignment.
- Staff positions and assist them in filling out their staffs.
- Ensure all team members are engaged in frequent communication and assist if normal job requirements conflict with team commitments.

Prepare and Mobilize
- Ensure individual and team readiness.
- Gather critical information pertinent to the assignment.
- Discuss team configuration and size with requesting agency.
- Establish a method of travel and team arrival time and location.
- Travel to the incident and monitor the team's mobilization status.
- Meet with the agency administrator to receive in-briefing, the delegation of authority or letter of expectation, and other pertinent incident information.

Build the Team
- Ensure team configuration, size, and qualifications are commensurate with incident complexity and meet the expectations of the requesting agency.
- Assemble appropriate team members and receive an in-brief of the current incident status.
- After the in-brief and team shadow period, establish a common operating picture with section chiefs.
- Establish effective relationships with relevant personnel.
- Provide training opportunities as appropriate.

Lead, Supervise, Direct
- Demonstrate leadership values and principles. Emphasize and monitor teamwork.
- Establish objectives; communicate priorities.
- Develop and implement plans and gain the concurrence of affected agencies and/or the public.
- Continually evaluate whether objectives are achievable given available resources and environmental, political, or socio-economic conditions.
- Monitor status and support staff.
- Identify political or agency issues that may impact the team or the incident.

Perform Incident Commander Duties

- Complete a daily review of the complexity analysis and incident objectives. Communicate status with the agency involved.
- Participate in any conference calls or meetings as established by agencies.
- Review and approve all incident reports.
- Schedule and attend community meetings as needed.
- Approve press releases.
- Consider smoke impacts to sensitive areas.
- Ensure incident financial accountability and expenditures meet agency standards.
- Ensure any transfer of command is effectively communicated.

Communicate and Coordinate, Manage Risk, Document the Event, Demobilize

- Demobilization plan. Brief assigned resources on demobilization procedures and responsibilities. Follow compliance with incident and agency demobilization procedures.
- Communicate and recommend demobilization time line to AA.
- Assist agency with complexity analysis promptly to facilitate an appropriate transfer of command.
- Return equipment and supplies to appropriate units and provide inventory and location of equipment and supplies to incoming IC.
- During the transfer of command:
  » Ensure continuity of operations.
  » Exchange critical safety information.
  » Communicate transfer of authority through the established chain of command.
- Anticipate potential resource needs for the incoming organization.
- Ensure turn-back standards provided by the hosting units/jurisdictions are completed and addressed.
- Ensure team members travel after appropriate rest and all C&G staff arrive safely at their regular duty station.

Figure 8. Incident commander responsibilities. (Summarized from National Wildfire Coordinating Group Incident Commander Job Description, https://www.nwcg.gov/positions/ict1#:~:text=The%20Incident%20Commander%20Type%201,Incident%20Action%20Plan%20(IAP.)

When a wildfire erupts, a local agency (city, county, or state) usually is the first to respond. The IC always is the highest-ranking officer on the scene in these initial attacks, regardless of training or experience fighting wildfires. The incident command function can transfer to more experienced personnel during a wildfire event depending on the fire's size and complexity. The IC position requires many years of training and extensive prior management and wildland fire management experience at the federal level.

> **Authors' Note**
> Immediate actions of the dispatcher and IC during the initial attack impact the outcome of wildfire events. The speed of response within the first few hours of outbreak fortified with the correct types and number of engines, firefighters, bulldozers, and aerial firefighting assets dictate the future size and duration of the incident. In the vast majority of instances, the event becomes either a fire event or a wildfire depending on the decisions made in those first few hours.

COMMAND AND CONTROL: THE ROLE OF THE INCIDENT COMMANDER TO ACCESS ASSETS

No matter the size of the fire, there are several questions that must be answered pertaining to available assets and tactical and strategic decision-making. The key questions are:

1. What happens tactically when emergency services receive a 911 call to report a wildfire?

2. What are the priorities for extinguishing, containing, or letting a wildfire burn?

3. Who or what activity makes that decision?

Federal law requires an IC, working closely with a dispatcher, is in charge of each wildfire. A call triggers a 911 response made to a centralized dispatcher, who then notifies the local fire department team. The most senior captain or firefighter deployed with this team usually is regarded or appointed as the IC. ICs and dispatchers can and will change if the wildfire becomes more extensive or complex.

The FS uses three classifications for command and control and asset management: initial attack, extended attack, and complex fire. The organizational command structure will change if or as the scope of the wildfire increases and assets available to the incident commander and dispatcher depend on the wildfire size.

Initial attack. This is the smallest of the responses and is triggered by wildfires 100 acres or less (slightly less than a quarter-square mile). In this phase, first actions may include assessment patrol, monitoring, hold or pursue an aggressive initial attack on a fire. The kind and number of resources responding to an Initial Attack vary depending upon the fire danger, fuel type, terrain, values to be protected, and such factors as distance to water sources and air attack bases. An initial attack uses local assets. The incident commander most likely will be from the first responding local fire department and the dispatcher from a local city or county centralized fire command center. Generally, the initial attack involves a small number of resources and a limited incident size—typically, a type 4 or type 5 rated fire (see fig. 9).

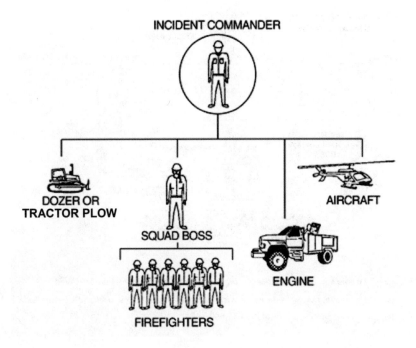

Figure 9. Example of initial attack organization (type 4 incident). (Figure by the National Wildfire Coordinating Group, https://www.nifc.gov/nicc/logistics/references/Wildland%20Fire%20Incident%20Management%20Field%20Guide.pdf.)

Extended attack. The second level response is for wildfires that grow or expand to a type 3 fire of about 300 acres (or about one-half square mile). Extended attack is the suppression or containment activity for a wildfire that has not been contained or controlled by an initial attack. Frequently, in this stage, a more experienced IC and operations team take over from the initial attack IC and dispatcher. The extended attack IC has additional equipment and personnel resources to combat the more extensive, more aggressive wildfire. The IC's staff also grows when a wildfire expands into the extended attack realm (see fig. 10).

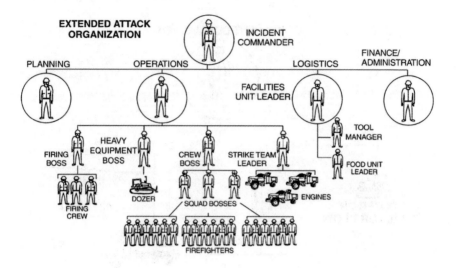

Figure 10. Example of Extended Attack organization. (Figure by the National Wildfire Coordinating Group, https://www.nifc.gov/nicc/logistics/references/Wildland%20Fire%20Incident%20Management%20Field%20Guide.pdf.)

Complex fires. The third response action is for wildfires covering more than 300 acres—typically, types 1 and 2. These fires require incident management teams headed by an experienced IC, who replaces the previously assigned ICs. These complex teams are prescreened, assembled, and staged to take over command of a wildfire exceeding the firefighting capabilities of the extended attack team. The complex fire team members are specially trained and prequalified personnel hired by either state or federal agencies, depending on the jurisdiction of the wildfire, to manage large-scale teams and complex wildfires. The teams may activate multiple fire agencies' and branches' resources, often managing 500 to 1,000 operational personnel, excluding actual firefighters. They are tasked with establishing active base camps and deploying mechanized and aviation assets of varying types and capabilities. Complex wildfire teams are assigned to the wildfire until it is extinguished (see fig. 11).

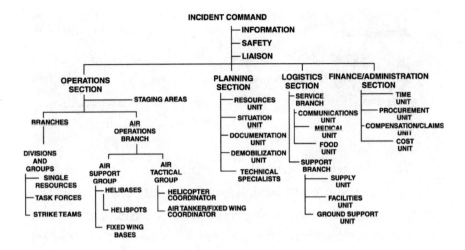

Figure 11. Organization chart for type 1 and type 2 incidents. (Figure by the National Wildfire Coordinating Group, https://www.nifc.gov/nicc/logistics/references/Wildland%20 Fire%20Incident%20Management%20Field%20Guide.pdf.)

TIME LINE ASSESSMENT

As noted by the GAO, when wildfires are not controlled during the initial attack (usually within the first twenty-four hours), they grow larger and more complex.[34] Data indicates that federal dispatchers rarely deploy mechanized equipment or fixed-winged aircraft during the initial response to a 911 call. On the other hand, county and state dispatchers, working with the IC, usually deploy all the assets available to them depending on their assessment of the wildfire conditions.

This difference in practice stems from the firefighting philosophies and mission statements of each local, state, and federal jurisdiction. Federal firefighting agencies have *not* adopted a "put the fire out" strategy in contrast with local and state jurisdictions that relentlessly pursue such practices. As discussed in chapter 2, federal multiagency memoranda of understanding (MOUs) require each federal firefighting agency to follow the agreed-upon guidelines on federal lands. For reasons unknown, putting out the fire is *not* listed as a priority in the latest MOU.

> **Authors' Note**
> Unless federal policy changes calling for a "put out the fire" mission statement, we can expect more extensive and more complex fires as the GAO discerningly concluded in its 2013 report.

The diagram below pictorially flowcharts a proposed cycle time for the initial response to a 911 call. It assumes the dispatcher is appropriately trained, particularly during the critical Initial Attack phase. This training includes informed decision-making to select the best assets based on the terrain and conditions (bulldozers, helicopters, aircraft) and engaging experienced crews using the equipment who understand the terrain and conditions (see fig 12).

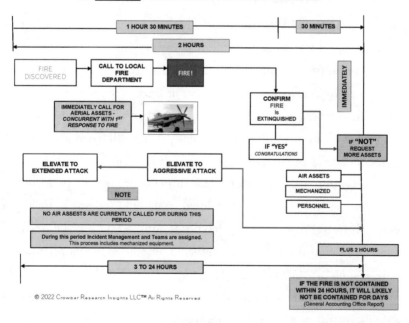

Figure 12. The proposed first twenty-four hours in the life of a wildfire. (Figure created by © Crowbar Research Design, LLC™. All rights reserved.)

Our proposed response time line follows many state practices; upon receiving the 911 call, immediately deploy single-engine attack tankers (SEAT) aircraft equipped with 800 to 1,200 pounds of water (or, in some instances, retardant). Activate the SEATs at the same time fire engines and crews are deployed. The cost of deploying these assets should be considered insurance against the risk that the initial fire becomes more dangerous, complex, and catastrophic. The relative cost of SEAT aircraft makes this choice the most responsive and economical. Fires up to 300 acres would require up to nine SEAT sorties (aircraft) to cover the fire line. If the SEATs are dispatched to attack the wildfire when the scope of the burn is 100 acres or less, only five SEAT sorties would most likely be required.

PREDICTIVE AND SATELLITE OBSERVATION SERVICES

NIFC, NICC, FS, and other federal, tribal, state, and local wildland fire management agencies use the Predictive Services program to assist in their strategic planning for wildland fires and management of resources. This program was created following the wildfire season in 2000 in association with the National Oceanic and Atmospheric Administration's (NOAA) National Weather Service (NWS). According to its annual operating plan, the NWS is responsible for "providing fire weather forecasts for the protection of life and property, promotion of firefighter safety, and stewardship over America's public lands," according to its annual operating plan.[35]

NWS uses various highly sophisticated forecasting tools, including Geostationary Operational Environmental Satellites (GOES) satellite thermal band technology, to predict, detect, and monitor precise wildfire incidence and behavior. These tools include specific longitude and latitude metrics for predicted wildfires and fire behavior metrics, such as intensity, movement, and rapid combustion indicators. Moreover,

NWS makes available to each federal GACCs, local, and state forestry agency monthly, seven-day, next-day, and spot forecasts of projected wildfire incidences. NWS also calls out "Red Flag" warnings using information compiled through its National Fire Danger Rating System. Based on the wind, humidity, fuel conditions (including fuel moisture levels), terrain/topography, and lightning risk, these forecasts indicate at least an 80 percent likelihood of wildfire eruption.

A specific Seven-Day Significant Fire Potential outlook report is produced by NICC using NWS forecasts during the fire season for predicting (with at least 80 percent probability) the significant wildfire potential for the next seven days. This report assists agencies in determining precisely where and when both state and nationally shared resources will be in demand in the United States in the next seven days.[36-39]

Recently, NASA's developers partnered with the Geospatial Technology and Applications Center (part of the FS) to create the Fire Information for Resource Management System (FIRMS U.S./Canada), a new and expanded wildfire mapping tool. It offers additional contextual layers and enhancements about wildfire incidents and classifies fires to show time since detection to depict active fire fronts, incident locations, and other information for current large fires in the United States and Canada (see fig. 13).

Figure 13. A screenshot of the FIRMS U.S. and Canada fire map highlighting the Creek Fire, which started on September 4, 2020, near Shaver Lake, California, as seen in the U.S. large incident locations layer. (Photograph by Earth Data—Open Access for Open Science, https://earthdata.nasa.gov/learn/articles/usfs-firms-us-canada.)

This tool provides active fire data within three to four hours of NASA and NOAA's satellites' observations. Wildland fire agency personnel can view imagery via interactive computer applications. In addition, it updates daily fire danger forecasts, current NWS fire weather watch, and red flag warning areas.

Recognizing improvements were needed to government agency predictive tools, private sector entities intervened, creating innovative solutions that play an essential role in predicting and detecting both wildfire incidence and behavior when a wildfire erupts. Remote Area Weather Stations (RAWS) and observation systems were deployed by NIFC, FS, NPS, and several states, including California and Texas. They substantially increased the predictability and behavior of wildfires. There are more than 2,200 RAWS deployed throughout the United States. These systems report real-time data with a system up-time of approximately 95 percent. Annual service agreements are in place with the private vendor to ensure timely software and firmware program updates and system performance.

RAWS units collect, store, and forward data to a computer system at the NIFC in Boise, Idaho, via GOES operated by NOAA. The data is forwarded automatically to several other computer systems, including the Weather Information Management System and the Western Regional Climate Center in Reno, Nevada.

RAWS monitoring devices provide real-time images, fuel conditions, wind speed and direction, air temperature, humidity, and soil moisture and temperature data. Portable RAWS are used selectively during wildfire occurrences for better "on-the-ground" data gathering; for instance, portable RAWS devices deployed during significant wildfire events provide valuable predictive information on wildfire direction, velocity, and underlying fuel conditions (see figs. 14 and 15).

Figures 14–15. Fixed Position FTS Remote Automated. (Above: Photograph by FTS, https://ftsinc.com/product/remote-automated-weather-station-raws/.) Remote Portable RAWS. (Below: Photography by FTS, https://www.nifc.gov/about-us/what-is-nifc/remote-automatic-weather-stations; used with permission from FTS.)

Combining NWS predictive systems with remote sensor devices offer firefighting agencies advanced notice of an impending wildfire risk caused by lightning with as much as 95 percent accuracy. With this information, firefighting assets can be staged, deployed, and immediately made available to contain and extinguish a wildfire before it becomes uncontrolled.

Today technologies exist to improve the predictability and the possible detection of wildfire occurrence and the size and potential outcome risk once a wildfire starts. FS immediate past Chief Vicki

Christiansen stated opportunities exist to optimize the use of technology during her testimony to the House Appropriations Committee on April 15, 2021. "We are also investing in several key technology and modernization portfolios, including Data Management, Enhanced Real-Time Operating Picture, Decision Support Applications, and Modern Tools for a Modern Response," Christiansen testified.[40]

ADVANCED BASES FOR WILDFIRE ASSETS AND AERIAL FIREFIGHTING AIRCRAFT

Each GACC, working with the NIFC and state agencies, is responsible for staging bulldozers, fire engines, aviation assets, and other fire equipment at regional staging areas in anticipation of wildfire events. The NICC and NIFC have secured 33 advanced aircraft bases in the fifteen most fire-prone states: twenty-three for single-engine air tankers (SEATs) and large air tankers (LATs), and an additional ten bases for very large aircraft tankers (VLATs). These advanced bases are designed, according to the FS, to reach most wildfires in thirty to forty-five minutes of flight time.[41]

Only California, Colorado, and Georgia coordinate aerial aircraft staging at the state level since the other most fire-prone states neither own nor lease any dedicated aerial firefighting helicopters or aircraft. The California Department of Forestry and Fire Protection (CalFire) has fourteen air attack and ten helitack bases throughout the state designed, according to CalFire, to reach most wildfires in twenty minutes of flight time.[42]

THE KEY TAKEAWAYS FROM THIS CHAPTER ARE:

- Fifteen states are the most fire-prone, and virtually all are located west of the Mississippi River.

- Lightning-caused wildfire incidents represent 65 percent of all wildfires in these fire-prone states, making them highly predictable.

- Human and utility power line–caused wildfires represent only 35 percent of all wildfires in the most fire-prone states.

- Each wildfire receives a complexity rating that guides the deployed assets and protocols to manage each wildfire incident.

- Dispatchers, incident commander, and incident command teams are critically important when promptly engaged and deployed.

- Also critical for wildfire control is immediate deployment of assets, particularly helicopters and SEATs, to control the wildfire within the first twenty-four hours.

- Prediction and detection systems are in place to predict wildfire incidence in advance with up to 95 percent accuracy, especially events with the threat of lightning.

- Increased use of the latest technologies can improve this predictability percentage, making each wildfire's duration shorter if firefighting assets are appropriately staged.

- Innovative solutions can further protect our forests and prevent wildland destruction if federal and state agencies embrace these opportunities.

- Advanced staging bases allow the positioning of firefighting equipment and aerial aircraft assets before a wildfire outbreak.

CHAPTER 5

PERSONNEL AND MECHANIZED ASSETS FOR FIGHTING WILDFIRES

Wildfire is a natural phenomenon that affects every corner of the country, from the wiregrass expanses of Florida to the New Jersey pine barrens to the rangelands of Texas.

—Tom Tidwell, former chief, USDA Forest Service[1]

Fighting wildfires requires a highly coordinated variety of resources. Theoretically, the incident command structure provides for the most effective deployment of personnel and other assets. During the initial 911 call, dispatchers determine whether the wildfire incident falls on state or federal lands. If the wildfire is on federal land, dispatchers can direct local, state, and federal resource response to the incident based on prenegotiated cost-sharing agreements.

If the right set of firefighting personnel or other assets are unavailable, or if the incident command teams lack information about the most appropriate personnel and assets to deploy, wildfire duration increases. This chapter focuses on the best personnel and land-based mechanical assets and prospects needed for fighting wildfires. Aerial aviation assets and the significant opportunities to improve their use are examined in the next chapter.

Table 1 includes various owned, contracted, and volunteer-sourced equipment for local, state, and federal firefighting efforts (see table 1). Pictured below are examples of some of the most impactful asset classes used today by the nation's firefighting agencies (see figs. 1–7).

Classification	Local and State	Federal
Firefighters	93,656(a)	10,000(b)
Fire Engines, Water Tenders	91,940	900
Bulldozers, Tractor Plow Units	2,851	210
Helicopters, SEATs, Light Aircraft	590	168
Air Tankers	13	18

Table 1. Summary of assets available to fight U.S. wildfires. a: total includes 10,430 seasonal employees; b: all seasonal employees. (Table created by Crowbar Research Insights LLC, from the following sources: Local and state information from National Association of State Foresters 2018 survey, https://www.stateforesters.org/where-we-stand/wildfire/; federal information from USDA Forest Service fiscal year 2022 budget justification- https://www.fs.usda.gov/sites/default/files/usfs-fy-2022-budget-justification.pdf.)

PERSONNEL AND MECHANIZED ASSETS FOR FIGHTING WILDFIRES

Figure 1. Large air tanker with retardant—Coulson B-737. (Photograph credit: Courtesy of and permission from Coulson Aviation USA.)

Figure 2. Single-engine attack tanker (SEATS); Air Spray Air Tractor 802 on floats. (Photograph credit: Courtesy of and permission from Air Spray USA.)

Figure 3. Helicopters with water snorkels. S-64 Air Crane™. (Photograph credit: Courtesy of and permission from Erickson, Inc.)

Figure 4. Bulldozer with air conditioning. (Copyright permission Mike Eliason, Santa Barbara County Fire Department.)

PERSONNEL AND MECHANIZED ASSETS FOR FIGHTING WILDFIRES

Figure 5. Firefighters on the ground. (Copyright permission Mike Eliason, Santa Barbara County Fire Department.)

Figure 6. Smoke jumper. (Photograph by the National Interagency Fire Center, public domain, https://www.nifc.gov/resources/firefighters/smokejumpers.)

Figure 7. Wildland firefighters. (Copyright permission Mike Eliason, Santa Barbara County Fire Department.)

PERSONNEL (FIREFIGHTERS, INCIDENT COMMAND TEAMS, SMOKE JUMPERS, AND HOTSHOT CREWS)

Local firefighters, whether full-time employed professionals or volunteers, are the first responders to virtually all wildfires. Local and state fire agencies require regular training, including simulation and continuing education to provide these personnel with up-to-date wildfire practices and procedures. Local teams are regularly used for federal wildfires. During preseason, individual firefighters from each local fire authority are evaluated and assigned to a team. They are assembled and deployed for two- to three-week intervals when called upon by federal incident commanders during the fire season.

Unless local and state firefighters have "red-card" certifications, however, they are not allowed to fight wildfires on federal lands. The red-card qualifications include physical fitness, experience, and training standards. This additional federal requirement can be a substantial impediment to securing firefighting personnel because it limits the potential number of local and state firefighters in the pool.

State agencies and the USDA Forest Service (FS) each employ about 10,000 firefighters during the fire season, generally from May through October, and are not retained beyond the fire season. The practice has drawn criticism because fire seasons are becoming longer. In early 2021, the FS committed to retaining 1,000 personnel from this workforce year-round to assist in off-season forest management. Unfortunately, significantly more personnel are required to clear forest underbrush and accumulated forest fuels than are engaged by the FS. From a statistical standpoint, 1,000 people cannot accomplish the FS's stated goals to improve forest management.

A more challenging issue in the past two years is the inability of the FS to hire an adequate number of federal firefighters to fight the nation's wildfires. The absence of a full complement requires more local and state personnel, which can delay deployment, increase costs, and delay extinguishing the wildfire. Despite increasing compensation for the 2022 fire season, significant hiring challenges continue. The 2022 fire season workforce of firefighting personnel and incident command team members were reduced considerably; recently, only 26,000 incident command team person-hours were staffed for 2022 compared to 35,000 hours for these teams in 2021, a 25 percent reduction. With the risk of wildfires increasing, a decline in vital incident command teams means longer duration wildfires, more smoke, and increased greenhouse gas emissions. It appears a combination of poor compensation offered to firefighters and incident command teams together with a decision to insource the hiring function from the previous long-term successful relationship with the Oregon Forestry Department all contributed to this crisis.[2]

Besides federal firefighting employees, the current practice includes additional deployment of military troops—and prison inmates for fires occurring on state land—to assist ground firefighting crews in complex fires. The training practices used to prepare these groups appear less than optimum. For example, it doesn't appear there are specific training protocols to ensure these individuals, particularly the military personnel, work safely alongside bulldozer operators and other firefighters. They often are provided with shovels and pickaxes and deployed to the wildfire site without adequate advance preparation.

Another major issue is the crew sizes for hotshot crews and smoke jumpers were reduced due to the inability to recruit personnel, most likely because of compensation issues. Shortages in these two groups can be catastrophic. Both are highly trained professionals deployed into complex wildfires and remote areas. Their work significantly bolsters the establishment of fire breaks while also coordinating aerial assets targeting drops of retardant and water.

MECHANIZED EQUIPMENT

A colloquial forestry definition of *mechanization* is a tool that protects the firefighter and leverages capabilities through machinery explicitly developed for tasks the firefighter cannot do efficiently by hand. Mechanization has a prominent but historically underutilized role in suppressing wildfires. The chainsaw, first used in 1926 in Germany, is the most widely used forestry tool since its invention, yet it took more than forty years to become a standard apparatus in forestry when in 1970, Robert McCulloch invented the eight-pound chainsaw.[3] Another device, the bulldozer, was developed for farming, but its blade was in use before the invention of any tractor. It consisted of a frame with a blade at the front, pulled by two harnessed mules. First used in the United States in 1920, it resembles the bulldozer currently used today.[4]

Today's sophisticated mechanized equipment includes forestry equipment with improved designs to increase operational safety and effectiveness in all phases of wildfire suppression and year-round forest management activities. Newer machine designs are replacing less efficient and more risky manual firefighter methods. Newer technologies enable equipment usage on more complex and steeper terrain. In most cases, these improvements in efficiency and safety also reduce environmental site impacts while helping to minimize suppression costs and losses to wildfire. (See the appendix for more information about today's heavy equipment assets.)

The key drivers to successful use of this equipment include

1. fire teams' experience with dispatched equipment,

2. knowing how to get the equipment to the wildfire site,

3. wasting no time on duplicative inspections,

4. understanding road conditions, and

5. lowboy truck access to staging areas inside the forest area.

Coordination responsibility relies on the combined efforts of the firefighting agency with jurisdiction, incident command team, Geographical Area Coordination Center (GACC) zone, and geographic area support personnel.

FIRE ENGINES AND WATER TENDERS

Fire engine and water tenders combine firefighter transport, water-carrying, and water-dispensing capabilities as the most identifiable heavy equipment. Federal agencies distinguish fire engine capability based on the following factors: engine capability, pump capacity, tank size, crew size in relationship to fire behavior, fire intensity, burning index, and rate of spread. Different classes of fire engines allow varying water tank and pumping capacities depending on the wildfire intensity. This equipment becomes a critical component of scheduled prescribed burns as a safety measure for the fire teams to clear forest floors.

BULLDOZERS

Bulldozers are the most prevalent heavy asset used in the forests today since they exponentially increase fire suppression efforts. Their goal is to create a fire line or fire break by removing and reducing the fuels from the forest as part of a wildfire effort. Dozers use three techniques in wildland firefighting: first—a direct attack, where the dozer is on the fire's edge; second—an indirect attack, where the dozer is in front of the fire, creating a fire break; third—a parallel attack, where the dozer can be anywhere from one to five feet in front of the flames, pushing the fire as the dozer moves. Some operators consider this the most dangerous aspect of a wildfire attack.

Figure 8. The Ponsse Scorpion Future Cabin. Manufactured by Ponsse, Rhinelander, WI. (Photograph by Ponsse, https://www.ponsse.com/en/web/guest/products/harvesters/product/-/p/scorpion_king#/.) [Photo credit: Ponsse, used with permission.]

Dozers also play a critical part in forest management and restoration projects. Significant bulldozer advances include tracked vehicles using weight dispersion to minimize forest floor damage, rubberized tracks, and advanced automation, all designed with forest care in mind (see fig. 8).

Each machine can have distinct features that improve or customize its capabilities. If appropriate features are not already factory installed, firefighting agencies and contractors can add aftermarket enhancements to improve fire suppression or forest management capabilities. This equipment blade is built to push soil or clear vegetation.

Bulldozers with similar features may perform differently depending on the operator's talent and experience, the quality of the road or trail, fire line locations, and the coordination with other suppression resources. Fire personnel require training and expertise to correctly assess their task objectives, bulldozer, equipment capabilities, and limitations. Experience facilitates assigning the correct tool for the right place and task. Accessing firefighting agency information—such as geographic information system-generated slope maps; aerial photos; tree stand data; and local knowledge of soils, terrain, and hazards—is critical in making informed equipment selections and geographic assignments. Several pieces of heavy equipment can be used in the various stages of the fire, including excavators for cleanup, establishing a fire line, and tearing out burned-out stumps or incinerated buildings. Excavators are used alongside dozers. Tracked shovels with the pushing power of a dozer can augment wildfire cleanup and scrape away soil layers to remove any earth that might contain burnt materials. They can scale steeper slopes and cross softer soil. A minor drawback is the logistics required to move the track loader by truck from site to site. Forest machines can be deployed with essential add-on equipment, such as felling heads, processing heads, and log and fork grapples, as well as water tanks. They are force multipliers for the dwindling number of firefighters on the ground and add significantly to productivity levels almost impossible for firefighter teams to meet.

Hiring and using local operators and equipment can provide non-local incident command teams with valuable information. Often local operators have modified their bulldozers to work best under local forest conditions. Their knowledge of ground and vegetation situations has proven invaluable, particularly in the lessons they've learned from current forest terrain, previous local fire incidents, and existing fuels treatments. Out-of-state incident command teams are unfamiliar with local terrains, making it a challenge to best use mechanized equipment. In these circumstances, localized equipment owners are extremely valuable in helping to choose the right and most available asset for the mission.

CONTRACTING, DEPLOYMENT, AND OPERATING PRACTICES

Federal agencies use standard procurement practices for large government organizations. These practices are time-consuming, filled with extensive paperwork, and designed in such a way that only existing government contractors likely succeed in securing new contracts. Mechanized equipment operators rarely receive firm commitments for specific equipment before fire season. Instead, most contractors are on a "call when needed" arrangement, meaning the equipment can be used on a wildfire only if the equipment is available and not in use on another project. This is not a desirable practice if the goal is to extinguish wildfires.

The following comments come from qualified bulldozer equipment operators about federal firefighting agency contracting and operating practices:

- "We used to do a lot of work with the Forest Service, but we don't do much for them anymore." When asked why, the response was, "They are tough to do business with and deal with."

- "Before awarding a contract, the Forest Service required us to deliver the equipment to their chosen site about 100 miles away for a prequalification inspection. Once we secured the contract, we had to have the same equipment inspected again before deployment, which was miles away from the wildfire, delaying our utilization by two days."

- "They don't know how to use the equipment delivered, such as feller bunchers."

- "They don't use the equipment for its intended use and efficiency."

- "People using the equipment don't know how to use it or supervise the best use in the circumstances." We asked for an example. "When firefighting crews encountered snags (i.e., a standing, dead or dying tree, often missing a top or most of the smaller branches), rather than use available equipment, crews were put in harm's way to establish a line (i.e., fire break)."

- "If a crew is developing a line, Forest Service policy requires our dozer to only cut to the beginning and from the end of the line on which the hand crews are working. It took the hand crews hours to cut down trees, stack logs, and remove slash to establish a seventy-five-foot fire line when the dozer could have completed the task in ten to fifteen minutes. Worse, once the fire approached the line, it broke through the line created by the hand crews, and the dozer-created lines held."

If the goal is to extinguish wildfires, the FS contracting process and practice are arcane, cumbersome, and irresponsible. Why? Because it lacks a sense of urgency and does not use equipment effectively. Continuing the status quo defies logical explanation.

Vendors that provide heavy equipment to fight wildfires receive contracts for one- to three-year terms covering the use of the equipment, fuel, crew costs, and transportation of the equipment to and from the wildfire site. Frequently a contractor can provide four or five different types of equipment under one contract with a "Task Force Agreement." A less common practice grants the commissioning of one piece of equipment.

Upon investigation, it became evident that dispatching, prepositioning, and training on heavy equipment severely lack effective management. Federal dispatchers frequently have little hands-on knowledge of equipment capabilities and the best use of what is available. It is difficult to determine dispatcher use of any regional contractor handy reference guide listing equipment, capabilities, and complementary equipment necessary for maximizing wildfire-fighting effectiveness. Our review indicates a possible FS bias against contracting with loggers in many regions. Logging firms and their crews utilize the best equipment and are the most knowledgeable about forest terrain in their areas. The better logging firms understand and employ sustainable forest management practices, which may dispel many environmentalist arguments.

The 210 dozers owned by the FS maintain 380,000 miles of forest roads, including those for fire access on federal lands.[5] If the bulldozers and its operators are available, they can be deployed to assist in fighting wildfires. Unfortunately, that is not their primary mission. Due to their limited numbers, equipment and their government-employed operators generally are not deployed away from their assigned forest areas even during fire season. Instead, contractors provide the heavy equipment and operators during fire seasons.

Equipment to be deployed for fighting wildfires must undergo three inspections: first—during-contract award; second—before deployment to a wildfire; third—after the wildfire is extinguished. The dispatch for the incident command team contacts the regional GACC, which notifies zone leaders who activate the prequalified contractor for wildfire deployment. There is virtually no prepositioning of the equipment because of the inspection requirement prior to deployment. Instead of mobilizing heavy equipment inspectors to the wildfire site, contractors must first travel to the FS inspection site, which could be hundreds of miles in the opposite direction. After completing a successful inspection, equipment and operators are deployed to the wildfire location.

The contractor then stages equipment and operators at the fire site and frequently serves as an outsourced dozer boss within the incident command team. Arrival at the wildfire site represents the first opportunity to determine if the correct equipment is in place and if all required equipment was assembled. Unfortunately, all too often, task force equipment does not meet all the wildfire needs. If time allows, the additional equipment is ordered, but too often some of the equipment and crews are never used.

Federal agency guidelines generally limit the number of hours crews can work to sixteen hours on and eight hours off. Machines can operate twenty-four hours a day with two crews. Federal regulations also contain inexplicitly different standards and operating procedures depending on the agency. For example, fire engine staffing requirements differ between the Bureau of Land Management, the National Park Service, and other federal agencies, while emergency lighting use regulations vary between the FS and other federal agencies. These federal regulations demonstrate how bureaucracy runs amuck, hampering the real mission to put the fire out. Yet federal agencies seem consumed with cumbersome rules that impede optimal wildfire suppression actions.

A PATH FORWARD

There is no question about it: Mechanization is underrepresented at the federal agency level and for most of the fifteen most fire-prone states. Given the labor shortage and dwindling numbers of firefighters and hotshot crews, mechanized equipment use takes on even more importance. Enhanced use and quicker deployment of mechanized equipment represent an untapped resource. Forests and properties at risk require more suppression resources, not less.

We have found no evidence of any "in-the-woods" hands-on, two-week training. The last formal training of this type related to western states—the most fire-prone area—was conducted in 2011. (The National Interagency Fire Center also conducted a day and night in-the-woods training session for its southern region at its Fire Training Center for the Mississippi Forestry Commission in 2019.) Without such training, dispatchers, incident commanders, and fire crews have little idea of the capability of their equipment, advances made by the vendor community to upgrade equipment capabilities, and current safety procedures when working alongside heavy equipment in daylight and nighttime conditions. Lack of regular annual hands-on training in a forest setting leads us to conclude that the federal agencies' strategy, mission, and core competencies require a serious reset. Just for safety alone, "in-woods training" should be a requirement, not simply an option.

Perhaps federal and state forest agencies should consider establishing a mechanized equipment charter with an appropriate command and control organization to prioritize training, deployment, cross-training, and safety practices. Much like other federal agencies, this division should hold vendor summits, where worldwide equipment manufacturers can demonstrate their latest technologies. State forest agencies, equipment contractors, and representatives from international fire and forest management agencies also could participate. Improved knowledge of equipment capabilities and feedback to manufacturers about desired future functions enhance both wildfire-fighting outcomes and forest management capabilities.

More important, as these vendor summits would demonstrate, much of the equipment for fighting today's wildfires can be deployed to thin forests, remove insect-infested trees, and slash and snags. Further, more efficient, economical, and environmentally friendly equipment is available because of recent technological advances.

These shortcomings in managing heavy equipment should give us all pause. Once again, commonsense business practices are nearly non-existent and the rules that are in place are disturbing. Some of the most productive, impactful assets to fight wildfires and manage our forests go mismanaged, unused, and underutilized.

THE KEY TAKEAWAYS FROM THIS CHAPTER ARE:

- Absent a full complement of appropriately hired and trained firefighters and incident command teams, wildfire risk and fire duration increased drastically.

- Prepositioning of bulldozers and various heavy equipment using the substantial investment in existing predictive systems requires immediate attention.

- A combination of lack of knowledge, poor planning, and an inadequate amount of heavy equipment assets directly impacts the ability to put the fires out.

- Additional heavy equipment could alleviate some of the firefighter labor shortages.

- Heavy equipment assets should be employed at night to fight wildfires, taking advantage of the lower temperatures, generally lighter winds, and higher humidity conditions.

- The contracting philosophy of the FS deters new entrants, stifles initiatives, and does not reward for updated and modern equipment.

- Vendor summits should be convened to take advantage of the latest heavy equipment technologies to fight wildfires and meet the critical need for forest management equipment.

CHAPTER 6

AVIATION ASSETS TO CONTAIN AND EXTINGUISH WILDFIRES

Not to have an adequate air force in the present state of the world is to compromise the foundations of national freedom and independence.

—Winston Churchill[1]

Historically, specialized helicopters and firefighting aircraft were used to fight wildland forest fires. These highly effective assets can

- protect homes, businesses, and real estate;

- extinguish the fire during the initial attack;

- slow an advancing fire by quickly establishing long fire lines (using environmentally-friendly chemical-based retardant to create fire breaks);

- develop a continuous fire break using retardant to gain control over the wildfire perimeters; and

- transport firefighters to battle the fire on the ground in remote areas.

Aerial firefighting assets deployment dates back to the late 1920s, when insecticide dropped from a Curtiss JN-6 airplane successfully treated catalpa trees. The first known aerial firefighting attack occurred in 1939, when a converted Stearman biplane crop duster dropped water on a grassland fire. Then, in 1953, as a test, a DC-7 prototype dropped its water-filled ballast contents over a mile-long swath during a low pass over the airport runway in Palm Springs, California. The first aerial drops on a forest fire occurred in 1955, when the Willows Flying Service used Stearman aircraft to drop 170 gallons on each flight to knock down hot spots on a Mendocino National Forest wildfire in California.[2]

After World War II, the USDA Forest Service (FS) experimented with military aircraft dropping water-filled bombs on forest fires. In the 1960s and 1970s, many former military platforms, including Lockheed's P-2 and P-3 aircraft, were retrofitted with gravity drop systems, including modular water and fire-retardant systems. Loaded into the fuselage, these retardant drop systems (RDS), also referred to as Modular Airborne Fire Fighting Systems (MAFFS), allow the aircraft to be used as an air tanker against wildfires. The United States fleet of air tankers currently uses improved versions of the MAFFS and RDS systems.

Today's federal fleet of aerial firefighting assets consists of 18 air tankers, 60 single-engine air tankers (SEATs), and 108 helicopters. This count excludes an additional "surge" fleet of aircraft and helicopters—8 MAFFS-equipped C-130s operated by the military and available to fight wildfires.

The U.S. air tanker fleet is comprised of former commercial passenger jet aircraft modified for firefighting use with retardant- and water-dropping systems capable of discharging their entire load of up to 9,400 gallons (from a DC-10) of fire retardant in less than twenty seconds, creating a fire break area one mile long by 100 feet wide. Air tankers can reduce the fire-retardant flow if the mission requires it. Through multiple drops, retardant-dropping aircraft can establish a perimeter around wildfires, removing the fire's fuel source and supporting the critically important fire crew working on the ground to contain the wildfire.

Effectiveness studies consistently demonstrate the value of aerial firefighting helicopters and fixed-winged aircraft. Since 2012, the FS, working with various universities, has conducted independent Aerial Firefighting Use and Effectiveness (AFUE) studies to assess the current fleet. Our analysis described in this chapter, in chapter 7, and in the appendix incorporates the March 2020 AFUE report findings.

Typically, cost and reimbursement for fighting wildfires are foremost in decision-makers' minds. Reimbursement factors include arrangements regarding the use of local (community or county), state, and federal funds. The federal government reimburses for firefighting on federal land, while state resources are responsible for state lands. The same policy, for the most part, exists for community fires. Local, state, and federal reimbursement agreements exist for wildfires that cross jurisdictions.

Incident commanders (IC) use aviation assets to monitor and suppress wildfires. These ICs also direct personnel and assets carried by airborne firefighting helicopters and fixed-wing aircraft to advantageous drop zones. Federal firefighting agencies' use of these assets, however, is not without controversy, as the following quotes illustrate:

- "There are advantages and disadvantages to each platform, and dropping on fires remains an art form, not a science."[3]

- "Wildfire suppression results in more fuel loads on the forest floors, creating more intense future wildfire incidents."[4]

- "Fire officials insist that fixed-wing air tankers are too vulnerable to the blinding smoke and high winds of extreme fire conditions . . . it would take nearly two hours for the first water-dropping helicopter to arrive and roughly six hours for the first air tankers to drop retardant on the fire."[5]

- Commenting on the size and effectiveness of the air tanker fleet in April 2021, the former head of the FS refused help for additional resources. "[W]e think we are really on the right track with our air tankers."[6]

HOW THE FLEET (HELICOPTERS AND FIXED-WING AIRCRAFT) IS ASSEMBLED TODAY

Third parties own virtually all the aerial firefighting assets at the federal level. The FS traditionally contracts with these independent operators to provide its fleet's aircraft, crews, maintenance, and logistics services. None of these third-party contracts extend beyond one year, although most agreements offer up to four one-year extensions. (The authors understand recent congressional consideration may soon extend the term of these agreements with third-party contractors by as much as ten years.) This contracting practice practically eliminates any incentive for firefighting companies to invest in the most efficient and effective fixed-wing aircraft, helicopters, or the technologies needed to enhance the assets they deploy. It also precludes or eliminates the opportunity for new entrants into the firefighting industry.

The aerial firefighting community represents a highly fragmented group of about ten to fifteen small- and medium-sized companies with limited access to capital required for significant investments in modern platforms absent long-term commitments from government agencies. Because contracts are limited to one-year terms, capital providers and financial institutions are reluctant to make substantial commitments.

In Canada's Manitoba province, the independent operator received a ten-year contract for seven water-scooping air tankers and three twin-engine commander "bird dog" aircraft.[7] In the United States, the Northern Rockies Coordination Center entered into multiyear contracts for excavators and fire engines for three-to-five-year periods.[8] Despite these precedents and years of complaints by the aerial firefighting contractor community, the one-year contract term practice continues to be FS policy. Congressional action is required to assist in revising financial models to accommodate multiyear contracting.

Conversely, local agency and state practices generally favor acquisition and ownership of heavy assets (fire engines, minimal numbers of helicopters, and fixed-wing aircraft), except for bulldozers, which usually are contracted through third parties. With owned assets, local and state agencies contract with third parties to provide the fleet's crews, logistics, and maintenance.

Under existing protocols, the FS estimates the threat for the upcoming fire season. Wildfire managers determine the expected number of aerial firefighting assets needed and issue solicitations to the firefighting industry. The FS uses two types of contracting instruments for aerial firefighting equipment:

- Essential use (EU) annual renewable contracts specify the equipment, crew requirements, and calendar days the equipment must be available, referred to as the mandatory available period (MAP). These contracts require 100 percent availability during the contract period, which is typically 160 days for large air tankers (LATs) and very large air tankers (VLATs) and 100 days for SEATs. Start dates are staggered to provide optimum continuous support throughout the fire season.

- Call-when-needed (CWN) contracts are awarded to companies for a "might-use asset." The company has no obligation to hold out that asset waiting for a FS call.

Complex solicitation and contract documents establish potential barriers to competition. Only existing contractors can navigate the impenetrable detail and expense involved in responding to and securing most contracts.

In airborne firefighting contracts, the EU and CWN contracts are assigned to companies by a specific air tanker tail number. Some contractors have aircraft on both EU and CWN contracts. The 160-day EU contract period (MAP) permits a contractor to have a plane on an EU contract for 160 days and a CWN contract outside the MAP period. The FS consistently makes public statements quoting the number of aircraft available to fight wildfires, including EU- and CWN-contracted aircraft, but that number can be deceiving. In actual practice, the figure could be less on any calendar day since an aircraft counted in the CWN contract could be committed simultaneously as an EU asset and therefore not available for CWN. Overstating the number of aerial firefighting aircraft available to fight wildfires seriously damages the institutional reputation and creditability of the FS and all federal firefighting agencies.

The second contract instrument, CWN, provides state and federal firefighting agencies with additional fixed-wing aircraft and helicopters should wildfires grow in number or intensity beyond the capability of EU assets. Use cost of a CWN asset can be greater than that of a like-type asset under the EU contract. According to Fire Aviation, "The 2017 average daily rate for large federal call-when-needed air tankers is 54 percent higher than aircraft on exclusive use contracts."[9]

Fire Aviation provided an analysis of the comparable costs of aircraft under an EU contract and a CWN contract: "We averaged the daily and hourly EU and CWN rates for three models of air tankers provided by three different companies, BAe-146 by Neptune, RJ85 by Aero Flite, and C-130 (382G) by Coulson."[10] The numbers below are the combined averages of the three aircraft:

EU daily rate: $30,150
EU hourly rate: $7,601
CWN daily rate: $46,341 (54 percent higher than the EU rate)
CWN hourly rate: $8,970 (18 percent higher than the EU rate)[11]

Air tanker use is expensive, so some ICs may be hesitant to deploy these assets in certain circumstances.

Unlike an EU contract, a provider with aircraft assigned under a CWN contract is not obligated to make that aircraft available to either state or federal agencies in any specific time frame. Accordingly, a CWN asset may *not* be available when needed. As a recent example, in 2020, the FS third-party contractor deployed one of its CWN LATs to Australia to assist in fighting its massive wildfires. At the same time, California was battling one of its largest-ever wildfires. As a result, despite the published number of airborne assets available to the public, this critical asset was not available to fight wildfires in the United States, specifically in California.[12]

Both EU and CWN contracts limit the number of consecutive days independent operators can use air tankers to six days on and one day off, with no night flying permitted for wildfire suppression. This practice supposedly allows light maintenance to be performed on the day off and, for safety reasons, with night flights, yet it obviously limits the maximum utilization of the contracted fleet. Light maintenance could be performed overnight, and night flight is possible with proper avionics and optical equipment. One day represents 15 percent additional use of critically important assets. Wildfires do not take days off, and neither should contracted air tankers.

Several states (California, Colorado, Kansas, Montana, Mississippi, and Georgia) own varying numbers of aerial assets instead of contracting through third parties. The most significant number is in California, with more than fifty dedicated air tankers and helicopters, making it the largest owned fleet in the world.[13] At the other end of the spectrum, Georgia has two SEAT aircraft and one helicopter capable of supporting ground attack personnel with water or fire retardant. Colorado and Montana also contract for LATs and VLATs to help those states' airborne firefighting asset availability.

Federal agencies established 33 advanced aircraft bases in the fifteen most fire-prone states: 23 for SEATs and LATs, and 10 bases for VLATs. Overall, there are 145 advanced bases located throughout the United States.[14] According to the National Interagency Fire Center (NIFC) and the National Interagency Coordination Center, these advanced aircraft bases permit the prepositioning of airborne assets to reach most wildfire-prone sites in thirty to forty-five minutes of flight time.

Authors' Note
Preparation time for an LAT aircraft and crew performing the regular safety and other preflight checks, and loading the water or retardant takes time. We believe the forty-five-minute published metric actually requires ninety minutes to two hours before an aerial fixed-wing aircraft can be engaged at the wildfire site. Awareness of this cycle time allows for better staging of the appropriate number of aircraft at each advanced base "to be on the ready" to establish a sustainable fire line.

COMPANIES WITH FIXED-WING AERIAL FIREFIGHTING ASSETS

All fixed-winged aircraft in the air tanker marketplace are retrofitted former commercial jet passenger aircraft, with the exception of the AT-802A Air Tractors, the CL-415 Scoopers, and the former military C-130s. Further, all planes except the AT-802A Air Tractor and the Canadair CL-415 Scooper were first manufactured more than twenty-five years ago (see table 1).

> **Authors' Note**
> Age may not necessarily indicate poor aircraft platform choice. Because routinely required maintenance practices must be followed, the engines, wings, and fuselage are replaced, rendering the aircraft age relatively less critical. What is essential is having the right aircraft platform for the firefighting mission and a readily available parts supply to replace items according to the aircraft's maintenance schedule.

Company Name	Headquarters
Dauntless Air (SEAT)	Appleton, MN
Aero Air LLC	Hillsboro, OR
Erickson Air (1)	Portland, OR
Aero Flite	Spokane, WA
Conair (2)	Abbotsford, British Columbia
Coulson Air (4)	Port Alberni, British Columbia
Neptune	Missoula, MT
Air Spray	Edmonton, Alberta, Canada and Chico, CA
10 Tanker Air Carrier	Albuquerque, NM
Global Supertanker (5)	Ceased Operations

Table 1. Aerial aircraft and contractors available to the Forest Service. In addition, there are about fifteen companies with SEATs in their inventories not listed. Footnotes: 1. Subsidiary of Aero Air; 2. Subsidiary of Aero Flite; 3. Conair DHC-8—currently under modification;

AVIATION ASSETS TO CONTAIN AND EXTINGUISH WILDFIRES

Inventory of Number of Aircraft Available	Name and Aircraft Type
15	802A - Air Tractors (wheels and floats)
2	MD-87
7	MD-87 and MD-88
4 6	CL-415 AVRO-RJ-85
2 2 6	AVRO-RJ-85 DeHavilland DHC-8T DeHavilland DHC-8 (3)
4 6 Numerous	Lockheed C-130 Boeing-737-300 Helicopters-Various Types
9	British Aerospace BAe-146
2 3 2 6 4 3	British Aerospace BAe-146 Lockheed L-188 (P-3 Military) 802F Air Tractor 802 Air Tractor Fire Boss Canadair CL-415 Scooper Canadair CL-215 Scooper
4	McDonnell Douglas DC-10-30
1	Boeing 747-400

4. Has a wholly owned subsidiary in the United States; 5. Ceased operations in May 2021. (Table created by Crowbar Research Insights LLC™, complied with data from Fire Aviation 2019, 2020, and 2021, and the AFUE report, © Crowbar Research Insights LLC™. All rights reserved.)

CASE STUDY—DEATH OF THE P-3: THE KILLING OF AN AERIAL FIREFIGHTING ASSET WITH A PROVEN HERITAGE

Sometimes research uncovers elements in case studies that appear to be senseless. That seems to be the case with the Lockheed P-3 Orion, formerly in service to fight wildfires for the FS.

The P-3 aircraft was NAVAIR-tested and designed to accomplish the critical mission of submarine hunting for the U.S. Navy.[1] The P-3 design factors perfectly fit the mission profile for aerial firefighting: it operated at low flight levels, made tight turns, stayed aloft (loiter) longer than any aircraft, and carried as much, if not more, water and/or retardant than any platform used by the FS today.

The P-3 finally ended its service life to the FS in 2011. The end had nothing to do with the aircraft's capabilities. The FS simply stated it no longer wanted to use old military aircraft to fight wildfires. The situation is more complicated. There was and is no substitute for the P-3 or its capabilities then and now. The decision to kill the P-3 platform had far-reaching, debilitating effects on the FS's ability to deploy and fight wildfires with LATs. The move immediately reduced the FS's national aerial firefighting capability by 50 percent. It's no coincidence that since 2000, the average number of acres destroyed each year by wildfires has increased more than 225 percent from the previous forty-year average. This decision was made despite the ranking of the P-3 as the best in its class to fight fires and the most economical solution available, according to a 2012 analysis.[3]

The preplanning for the removal of the P-3 may well be seen as shortsighted. In 2012 and 2013, attempts to resurrect the P-3 for national aerial firefighting were inexplicably and summarily dismissed. Yet there could have been a solution—the FS could have requested aircraft manufacturers' designs for an aircraft used explicitly for aerial firefighting if the agency had some foresight.

For many years, the P-3 aircraft was designed to take the punishment of high g-force operations—a necessity in close-cover wildfire operations. One of the best examples of aircraft use in close cover of firefighters on the ground at altitudes of less than 150 feet above ground level (AGL) occurred near Boulder, Colorado, in 2010.[4,5]

This case study leads to several questions: Why is there a lack of understanding of matching aircraft mission requirements to aerial firefighting? There is a growing need for additional specially designed aircraft and a system for aerial assets on-site of a wildfire within ninety minutes instead of a current general practice of two to four days. Why is this still unresolved, with no clearly published plan for aerial-asset use?

Currently approved airframes for fighting wildfires are not matched to the mission. They are constrained by their ineffective airframe capability to carry higher volumes of water and retardant, inconsistent water and retardant drop accuracy, inability to operate at low altitudes, and maneuverability when faced with challenging wind and difficult terrain conditions. The current fleet is expensive and not designed to provide optimum service to the FS or citizens of the United States.

The aircraft in use today were designed to commercial airline specifications and never intended for dropping a required amount of retardant or water on fire. The current Boeing 737-300 series derivative aircraft in use for the FS was also designed for airline and commercial usage and not low-level aerial firefighting. That airframe is stressed to +2.5 and −1 g-force, not to the more resilient former P-3 Orion standards mentioned above.

For more than fifteen years, FS request for proposals were limited to a suboptimum aerial platform for aerial firefighting.[6] Retardant and water capability (both in volume carried and accuracy of drops) and maneuverability limit aircraft use and design. We must question why airborne tanker design standards have not been issued matching the FS needs beyond what is minimally required.

Footnotes

1. Airframe designed to withstand +3 and −1 g-forces—a mission standard applied to the P-3.

2. Except for the DC-10; or B747 aircraft that the FS has now grounded.

3. "Airtankers and wildfire management in the U.S. Forest Service: examining data availability and exploring usage and cost trends," U.S. Forest Service, last modified August 10, 2012. https://www.fs.usda.gov/treesearch/pubs/41329.

4. The Gold Hill Fire, Boulder, Colorado, September 6, 2010.

5. Aero Union tail number 22; reported approximately <150' AGL (above ground level). www.youtube.com/watch?v=iMU5soCRY.

6. NEXGEN Solicitation 12024B18R9013 dated December 3, 2018.

HELICOPTERS

Helicopters have a unique place in the firefighting system. Rotor-wing aircraft delivers an impressive attack on wildfires, primarily with water, though most helicopter delivery systems are capable of attacking ground fires with retardant. They can recover and refill water from local bodies of water inaccessible to traditional fixed-wing aircraft and operate with more rapid return cycles to the hot spots. Helicopters load from water sources close to the area of operations. The wildlands and forests often have water sources (streams, rivers, and lakes) nearby for replenishment and even swimming pools when the need arises. The latest AFUE report states helicopters achieve the best outcomes, especially in initial attack, with up to an 87 percent effectiveness in specific drop zones.[15]

For the 2021 season, the FS, in addition to fixed-wing aircraft, had at its disposal 108 EU helicopters and about 200 CWN helicopters.[16] Similar to bulldozers and fixed-winged aircraft, helicopters are classified into three categories, depending on their size and capabilities (see figs. 1–3).

Figure 1. Category 1: S-64 Air Crane™. (Photograph credit: Courtesy of and permission from Erickson, Inc.)

Figure 2. Category 2: UH-1H "Super Huey" Type 11 helicopter. (Photography by CAL FIRE, 2019, https://www.fire.ca.gov/media/4950/aviation-guide-2019-access.pdf.)

Figure 3. Category 3: Bell 206 Helicopter. (Photograph by Los Angeles City Fire Department.) 206A (Courtesy of LAFD – permission requested.)

Category 1	Large-capacity helicopters capable of lifting 700 to 3,000 gallons of water or retardant. Twenty-eight EU heavyweight helicopters were available in the 2021 fire season, including Kaman KMax; Boeing's CH-46, CH-47, and CH-54; and Sikorsky's S-61, S-64, and S-70.
Category 2	Medium-capacity helicopters with a water- and retardant-carrying capacity of 300 to 400 gallons. Thirty-four helicopters in this category, mainly the Bell 205 and the Bell 212, operated in the United States during the 2021 fire season.
Category 3	The lightest helicopters, with a payload of 100 to 200 gallons of extinguishing products: water or retardant. Forty-six EU helicopters were available in 2021.

Figure 4. A flight crew from Multinational Battlegroup East fills a Bambi Bucket. (Image by Joshua Dobbs/U.S. Army, public domain, https://www.donarney.com/news-blog/2018/1/8/don-arneys-bambi-bucket-a-major-contribution-to-aerial-firefighting-tech-from-the-1980s.)

Figure 5. Bambi Bucket. (Image with permission and from SEI Industries, https://www.sei-ind.com/.)

Helicopters equipped with snorkel hoses or Bambi Buckets can access lakes and water sources near a wildfire enabling frequent rotations against a wildfire. These helicopters also can replenish water used by heavy equipment on the ground, such as water tenders, skidgines, and forwarders. Invented in 1982, the Bambi Bucket is collapsible—some versions fold into the size of a set of golf clubs—easy to install on a standard helicopter cargo hook, easy to fill, and precise. It is capable of releasing a column of water directly onto a fire (see figs. 4 and 5).

Recently, three companies—Helimax, Columbia Helicopters, and Billings—acquired and now use retrofitted internal bunkers allowing their CH-46 and CH-47 helicopters to operate at higher intervention speeds while taking up to 3,200 gallons of water internally. As a result of these modifications, the internal capacity for these helicopters is greater than many of the LATs.

In 2021 Coulson Aviation contracted with California's Orange County Fire Authority for night-flying helicopter operations with a Chinook CH-47 platform. This novel aerial asset provides up to 3,200 gallons of water and retardant capacity, state-of-the-art avionics, and enhanced nighttime vision and communication equipment for conducting safe nighttime engagement of wildfires. Southern California Edison contributed $2.2 million toward the annual lease cost of this asset to protect their customers and critical infrastructure.[17]

Unfortunately, the poor administrative practices at the federal level can negatively impact operational assets during fire season. For example, the FS was so late in letting contracts for helicopters that *no new contracts* were issued for the 2022 fire season.[18] These FS administrative failures received little mainstream media scrutiny, allowing the FS to act without consequence while putting our national forests at additional risk and the lives and livelihood of nearby citizens in jeopardy.

FIXED-WING AIRCRAFT

Within the fixed-wing classification are three subcategories of aircraft: SEATs, LATs, and VLATs. These ratings are primarily based on the aircraft's water- and retardant-carrying capabilities and the number of pilots required to operate the aircraft:

SEATS

- Maximum of 800 gallons of water or retardant

  » Air Tractor 802, either on wheels or floats

LATS

- Type 1: 3,000 to 5,000 gallons of retardant.

  » MD87/88s, the C-130s, the 737s, BAe-146s, and the RJ85s

- Type 2: 1,800 to 2,999 gallons of retardant

  » DeHavilland DHC-8/400T (not currently in use but being modified by Conair)

- Type 3: 800 to 1,799 gallons of retardant/water

  » Grumman S-2T

VLATS

- 8,000 gallons or more of retardant

  » DC-10 (and potentially the return of the Boeing 747).

SEATS

SEATs can be a single- or twin-engine aircraft flown by a single pilot. SEATs are used for rapid initial attack delivery of water or fire retardant on wildland fires. These air tankers are strategically located to respond to the most remote areas within twenty to forty-five minutes of flight time and to a range of 500 miles (see figs. 6–7).

Figure 6. Grumman S-2T Type III air tanker. (Copyright permission Mike Eliason, Santa Barbara County Fire Department.)

SEATs' water and retardant capacity of 800 to 1,200 gallons makes them a key component of many nations' wildfire suppression strategies, particularly during initial attack.[19] Because these aircraft deploy rapidly, drop water payloads accurately, and return to base for another rotation quickly, SEATs can extinguish fires during an initial attack if swiftly deployed.

SEATs represent a very versatile asset, capable of quick response. According to Air Tractor, the Air Tractor can deliver up to 3,200 gallons of water or retardant an hour.[20] The Scooper model of the AT-802 Fire

Boss can increase that delivery capacity to almost 13,000 gallons of water per hour depending on the location of the nearest water source.

Figure 7. Air Spray Air Tractor 802A on floats (the aircraft also is available as a land option). [Photograph credit: Courtesy and permission of Air Spray USA.]

The California Department of Forestry and Fire Protection (CAL FIRE) owns 23 Grumman S-2Ts twin-engine SEATs with 1,200-gallon retardant capacity tanks. Colorado also maintains two SEATs under EU contracts, each with 800-gallon water or retardant capacity as part of its total fixed-wing fleet of three aerial firefighting aircraft. Kansas recently acquired two S-2 Grumman aircraft for its airborne firefighting support. Georgia and North Carolina maintain two Air Tractor SEATs aircraft for air attacks on local and state fires.

The Bureau of Land Management (BLM) is responsible for management of the federal-level SEAT program. In 2022 BLM had 60 EU SEAT contracts and 55 SEATs under CWN contracts.[21] SEATs rank just behind the helicopter in terms of drop accuracy—the metric that tracks placement of water or retardant on the correct coordinates. This asset represents the lowest cost and most economical platform in the FS fleet arsenal.

LATS AND VLATS

LATs and VLATs represent the highest retardant capacity aircraft platforms. They are typically engaged by IC teams during either extended attack or complex attacks and rarely are they deployed during initial attack. Unlike SEATs or helicopters, the current fleet of LATs and VLATs are all converted commercial aircraft (see figs. 8–11).

Figure 8. Neptune BAe-146 large air tanker. (Photo © John Hall, Courtesy of Neptune Aviation.)

Figure 9. Erickson Aero Tanker MD-87. (Photograph credit: © Courtesy of Marty Wolin and Erickson Aero Tanker LLC.)

Figure 10. Coulson Aviation C-130H. (Photograph credit: Courtesy and permission of Coulson Aviation USA.)

Figure 11. 10 Tanker DC-10-30. (Photograph credit: Courtesy and permissions of 10 Tanker Air Carrier.)

The table below describes the LAT and VLAT aircraft under contract with the FS. The lift capacity of each platform for this fleet from 2019 to 2021 is included. (see table 2).

Aircraft Type	EU 2021	CWN 2021	EU 2020	CWN 2020	EU 2019	Lift Capacity in Gallons
DC-10-30 VLAT	2		2	1	2	9,400
C-130 Series	2		1		1	4,000
B-737		1		1		4,000
BAe-146	4	1	4	5		3,000
MD-87	4	1	2	4	2	3,000
RJ-85	6	1	4	2	4	3,000
CL-415 Scooper					2*	1.600
Totals	18	4	13	13	11*	

*Note: In 2019 the CL-415 contract was canceled due to budget cuts.

Table 2. LAT and VLAT aircraft under contract with the FS. (Table created by Crowbar Research Insights LLC™ complied with information from Fire Aviation March 8, 2021, © Crowbar Research Insights LLC™. All rights reserved.)

CASE STUDY—THE BOEING-747 SAGA

The Boeing 747 can carry 19,200 gallons of fire retardant. During our research and development of this book, the aircraft was grounded due to inconsistent constant flow of retardant from its automatic retardant drop system. The aircraft was eventually sold, the retardant-dropping system fixed, and the aircraft readied for aerial firefighting action. The new owners projected the B-747 would be available for wildfire firefighting before the end of the 2022 fire season. It now appears that will not be the case as Alterna Capital Partners LLC, owners of the only Boeing 747 retrofitted for firefighting use, grounded the aircraft, basing the decision on failure to generate enough profit from firefighting contracts. According to Roger Miller, managing director of Alterna Capital Partners LLC, "The technology investment to upgrade the drop . . . system and the 747-400 aircraft was very significant. The economics under the company's current federal 'call when needed' contract are such that the investment will take longer to realize profitability than previously expected."[22]

Based on our analysis, the B-747 demonstrated the lowest cost and highest drop accuracy of any LAT or VLAT aircraft in the FS arsenal. This unfortunate development means the 2022 fire season will be without this important tool. The aircraft was sold again and is not expected return to service.[23] The B-747 has the largest retardant capacity of any firefighting aircraft at 19,200 gallons, ideally suited to establish fire lines for large, forested areas, expansive grasslands, and savannahs. Despite the aircraft owner's best efforts and significant financial investment, the FS reward of only a CWN contract defies reasonable expectations or thoughtful alignment of fleet strategy.

It appears the FS cannot manage large air tanker acquisitions effectively. The case study below describes the Lockheed Hercules C-130 transaction. Like the B-747 example, this episode demonstrates that the FS does not have the internal expertise to select and manage a VLAT fleet strategy.

CASE STUDY—THE C-130 HERCULES FIASCO

In December 2013, President Barack Obama signed the 2014 National Defense Authorization Act directing the United States Coast Guard to transfer seven Lockheed C-130 Hercules aircraft to the FS for wildfire suppression and other uses. The legislation also appropriated $130 million to convert these planes into air tankers and perform the maintenance needed for this fleet.

One would expect the FS to be elated to own this fleet to bolster its wildfire suppression mission while also adding significant personnel and materials transport capacity. This was not the case. Instead, on June 1, 2015, the FS distributed a briefing paper stating it was not prepared to take on the safety oversight that would require the agency to hire two full-time safety personnel explicitly assigned to this fleet.

Instead of taking advantage of the time after the award (December 2013) to create these positions and absorb the additional fleet, the FS waited another 18 months to make this decision. Its briefing paper stated, "the time frame to create one or more new positions to provide aviation safety oversight duties would likely be lengthy and not meet the agency C130H requirements in time for the 2015 fire season." In the briefing paper's conclusion appeared this eyebrow-raising comment: "This is a new program for the Forest Service, one that we have never managed before (we don't know what we don't know."[24]

The FS decided it did not want the air tankers more than four years later. In 2018 Congress passed additional legislation transferring this fleet to California for use in fighting state wildfires and increased the 2013 maintenance appropriation from $130 million to $150 million.

California wildfires historically are the most intense and extensive requiring maximum use of all available assets. Rather than immediately delivering these planes to the CAL FIRE, California requested the United States Air Force complete the maintenance and retrofitting required to qualify them as an air tanker fleet.

CAL FIRE now expects two of the C-130H fleet to be ready for the 2022 fire season and the remaining five in 2023. Almost ten years will have elapsed before desperately needed assets are fully deployed. This is anything but a rapid response and a job well done. The case study is emblematic of bureaucratic ineptitude and mismanagement.

WATER AND FIRE RETARDANTS

WATER

Water is highly effective in the initial stages of a wildfire incident. Whether delivered by helicopter or SEAT, its early use can extinguish most fires *if* these assets are deployed rapidly during the initial attack. Forests often contain nearby streams, ponds, and lakes that state and federal firefighters along with airborne assets—helicopters and fixed-wing scoopers—use to replenish the water for each drop. Because each wildfire is different, the availability of water sources for access by helicopters and fixed-wing scoopers for fire suppression is not always an option.

Most studies indicate successful wildfire suppression and extinguishment occurs during an initial attack using water-carrying fire suppression equipment such as helicopters and SEATs; that is *only if caught early*. Helicopter drops are highly effective when water sources are located within twenty minutes of the blaze (between thirty and forty miles).[25] Helicopters with snorkels or Bambi Buckets can replenish water-carrying heavy equipment—that is, fire engines and skidgines. Helicopter rotation or return times are generally shorter when using local water sources than for airport-based fixed-winged aircraft using advanced bases.

On the other hand, water use at the height of a wildfire incident can only slow down the intensity of most wildfires and lower the ambient temperatures. Several studies indicate that up to 85 percent of water dropped during an intense fire evaporates.[26] Photographs by mainstream media showing helicopter water drops attacking a wildfire might reassure the public, but remember the purpose of these events is to reduce the wildfire intensity only, not extinguish it.

FIRE RETARDANT

LATs and VLATs rarely use water during wildfire suppression operations. From the FS contract with air tanker operators: "The primary mission for air tankers under this contract is dropping retardant on or near wildland fires. Loading or dropping water shall not occur unless previously approved by the Forest Service National Airtanker Program Manager."[27]

Phos-Chek fire retardant, the primary fire-retardant chemical for wildfire suppression, was first commercialized in 1963. Developed by Monsanto, the powder mixture was later approved by the FS. The brand name Phos-Chek stems from its active ingredient, ammonium phosphate. Application of Phos-Chek creates a barrier or fire break to check and stop the spread of a wildfire—*it does not put the fire out*.

Unlike water or firefighting foam, fire retardant is *not* meant to be applied to burning fuels. Instead, it is used on vegetation surrounding a wildfire event to create a chemical fire break by depriving the oncoming fire of fuel. Once the red slurry douses vegetation, the affected plants will not incinerate. Further, the foliage remains nonflammable until a heavy rain washes off the mixture.

Fire retardant is generally safe. The FS states the risk of chemical toxicity is minor for most animals and predicts no risk for accidentally splashed firefighters during an aerial retardant drop; however, the FS advises care around bodies of water and streams due to possible minor impacts on fish and water-centered wildlife.

Perimeter Solutions, Inc., headquartered in Saint Louis, Missouri, is the sole-source supplier of Phos-Chek to state, federal, and several international agencies responsible for wildfire suppression. EverArc Holdings, Ltd., acquired Perimeter Solutions, Inc., in the summer of 2021.

Phos-Chek generally is mixed at the advanced aircraft staging bases into a slurry and is uploaded into aerial fighting aircraft using pressurized hoses. Depending on the tank capacity of the aerial firefighting aircraft, a Phos-Chek recharging can take anywhere from eight to twenty minutes.

LAT and VLAT aerial firefighting aircraft lay down an uninterrupted "chain" of retardant to create a chemical fire break from an oncoming wildfire. A chain is defined as 66 feet; 80 chains equal one mile of fire control line; and 10 square chains equal oneacres.[28, 29]

Depending on the terrain and fuel source, 100 gallons of retardant create one chain for short grass, perennial grass, or western woody shrub. The same 100 gallons make 0.6 chains of fire line on grass, pine, sawgrass, tundra, or pine litter or 0.4 chains of fire line on all other fuels, including hardwoods, pine, and slash.[30] Therefore, the IC must immediately assess terrain and fuel source as part of the initial attack. Generally, 8,000 gallons of retardant are needed to establish one mile of fire line.

The table below summarizes Phos-Chek's use over the past 25 years (see table 3).

Phos-Chek	Gallons Deployed	Cost in 2021 Dollars	Acres Destroyed by Wildfires
1994	22 million	$69 million	4 million
2020	56 million	$175 million	10 million

Table 3. Use of Phos-Chek over the past twenty-five years. (Table created by Crowbar Research Insights LLC™ compiled with information from 2020 USDA Aviation Annual Report and NIFC annual wildfire statistics, © Crowbar Research Insights LLC™. All rights reserved.)

Other fluids and chemicals and retardants, including foams and gels, can effectively contain wildfires. Unfortunately, the systems used for rapid refilling, dropping, and distribution have yet to be commercially viable.

Authors' Note

While we have no evidence of any concerns about the contracts with Perimeter Solutions, sole-source contracts represent an opportunity to self-select expensive options due to the lack of a competitive bidding process. By limiting competition and not rebidding the contract in regular periodic cycles, opportunities to enhance novel, more cost-effective and efficient solutions developed in the marketplace are stifled. Furthermore, sole-source contracting potentially exposes the supply of Phos-Chek in the event of production interruptions or manufacturing shutdowns by the supplier. We strongly discourage sole-source contracting for these reasons.

THE KEY TAKEAWAYS FROM THIS CHAPTER ARE:

- The equipment solicitation process is too narrow and prescriptive and does not provide the level of specifications expected to implement an optimal fleet strategy.

- The launch window for LATs and VLATs must be adjusted to ensure assets are on the fire within several hours instead of several days.

- The annual contracting process must be improved to ensure an adequate fleet is available before the start of each fire season.

- The EU contracting for MAP periods of 160 days requires adjustment to a minimum of 270 days (nine months) to recognize the longer fire seasons.

- Use of aircraft only six of seven days means 15 percent of fleet capacity is not available during the height of fire season. Wildfires do not take days off, nor should the air tanker assets available to fight them.

- Mandatory days off scheduled months in advance have proven untenable. Duty and rest periods should be matched to previously accepted regulations to provide flexibility and efficiency (thirty-six hours in seven days and one twenty-four-hour break in seven days).

- As demonstrated by California's Orange County Fire Authority with its acquisition of a night-flying, retardant-dropping helicopter, examine options to equip the air tanker fleet with avionics and optical systems to allow night wildfire suppression flights.

- Currently, only one retardant chemical is used by the FS to contain wildfires. There are two issues: (1) a sole-source supplier of a critical chemical potentially could restrict the supply of retardant due to production interruptions in the midst of fire season; and (2) what are the FS initiatives to foster competition among other companies to develop lighter-weight, potentially more effective products?

CHAPTER 7

COSTS AND BENEFITS OF TIMELY DEPLOYMENT OF ASSETS

The ultimate authority must always rest with the individual's own reason and critical analysis.

—*Dalai Lama*[1]

Government firefighting agencies provide significant information about aerial wildfire suppression costs in their annual reports and appropriations requests. The information, media coverage, and our own research developed the content contained in this chapter.

CURRENT COST ANALYSIS

SINGLE-ENGINE AIR TANKER (SEAT) ANNUAL CONTRACTED COSTS

According to the Department of the Interior 2021 annual budget, its Office of Wildland Fire spent $384 million in 2021 and an equal amount in 2020 for wildfire suppression and response.[2] Costs include

SEATs, helicopters, bulldozers, firefighters, and incident command personnel engaged in wildfires within Bureau of Land Management jurisdictions. The budgets do not report the expected contracted costs for SEATs; however, we estimate them between $110 million to $130 million annually.

HELICOPTER ANNUAL CONTRACTED COSTS

In 2020 helicopter expenditures were $402.4 million, significantly greater than the five-year average of $335.5 million (see table 1). More than 60 percent of costs were attributed to availability charges for helicopters on standby awaiting calls into service.

Calendar Year	Flight Hours	Availability Costs (Millions USD)	Flight & Other Costs (Millions USD)	Total Costs (Millions USD)
2016	34,371	$173.6	$208.4	$382.0
2017	43,981	$200.2	$144.5	$344.7
2018	39,892	$198.0	$141.0	$339.0
2019	20,588	$145.2	$64.3	$209.5
2020	42,667	$251.0	$151.4	$402.4
5-year average	36,300	$193.6	$141.9	$335.5

Table 1. CY 2016–2020 Contract helicopter use and costs summary. (Table from USDA Forest Service 2020 Aviation Annual Report, January 1, 2021, https://www.fs.usda.gov/sites/default/files/2021-06/CY2020_USFSAviationReport_Final_1.pdf.)

COSTS AND BENEFITS OF TIMELY DEPLOYMENT OF ASSETS

> **Authors' Note**
>
> If 2019 costs were deleted from the table as an unusual year, the average cost for the remaining four years is $367 million. Using the "adjusted" average, the 2020 increase was approximately 10 percent greater than this revised average.

LARGE AIR TANKER (LAT) AND VERY LARGE AIR TANKER (VLAT) CONTRACTED COST DATA

The below figure from the USDA Forest Service (FS) 2020 Aviation Annual Report summarizes payments made to aerial firefighting companies in that calendar year for LATs and VLATs (see fig. 1).

Figure 9 – CY 2020 Contract LAT/VLAT Costs by Pay Code Description

Pay Code	Total Costs
Availability	$123,114,127
Flight Time	$81,684,986
Extended Standby	$1,318,980
Other	$472,170
Total	$206,590,262

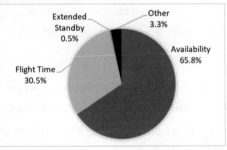

Figure 1. CY 2020 contract LAT and VLAT costs by pay code description. Contract LAT and VAT expenditures were $206.6 million in CY 2020. Availability charges total $123.1 million (65.8 percent) and flight time costs were $81.7 million (30.5 percent). Standby and other expenses accounted for the remaining costs ($1.8 million, 3.8 percent). All LAT and VLAT expenditures (not including scoopers) represented 27.9 percent of total agency aircraft expenditures billed in the Aviation Business System. (Table from USDA Forest Service 2020 Aviation Annual Report, January 1, 2021, https://www.fs.usda.gov/sites/default/files/2021-06/CY2020_USFSAviationReport_Final_1.pdf.)

SUMMARY

We used the information from each set of cost data above for each type of aerial firefighting asset to tabulate the average cost in the table below (see table 2).

Aircraft Type	Total Annual Cost in 2020	The Number under Contract in 2020	Average Cost for Each Asset
SEATs	$110 to $130 million	60	$1.8 to $2.2 million
Helicopters	$402 million	108	$3.7 million
LATs/VLATs	$207 million	13	$15.9 million
Total Cost in 2020	$719 to $739 million		

Table 2. Average costs for wildfire aircraft assets. (Table created by Crowbar Research Insights LLC™ compiled with information from USDA Forest Service 2020 Aviation Annual Report, © Crowbar Research Insights LLC™. All rights reserved.)

As demonstrated by table 2, wildfire aircraft assets are expensive. We identify several opportunities to significantly reduce this annual cost in this chapter. One cost-saving opportunity involves potential use of the military fuel supply system, enabling the purchase of fuel at military negotiated rates instead of more expensive market prices. Actual retail fuel costs reimbursed to the independent operators are included in the tables above.

PREPOSITIONING, CYCLE TIMES, AND OPERATING PRACTICES

The 2012 Rand Corporation *Air Attack Against Wildfires* report estimates that a successful initial attack can save, on average, $2.1 million per incident ($4.3 million in 2022 dollars) by avoiding the incremental cost of an expensive larger wildfire.[3] Applying this metric to the fifteen most fire-prone states' average of 2,752 yearly lightning-caused wildfire incidents greater than 300 acres creates an opportunity to avoid almost $12 billion per year in economic losses.[4] This savings increases to $20 billion annually if human-caused incidents are included.

Success in extinguishing a wildfire depends on the arrival of the first air tanker, a finding confirmed in a 2020 study presented in the following table. Note that 10,613 of the 11,665 fires (91 percent) were extinguished in under two days when air tankers were engaged within the first six hours of the initial 911 call (see table 3).

Time Elapsed Between Initial Reporting and Arrival of Airtanker (HOURS)	Fires	Fire Duration (DAYS)		
		Median	Mean	90th percentile
<1	8,234	.02	.15	.06
1-3	2,156	.11	.67	1.08
4-6	223	.25	.85	1.91
7-12	152	.56	1.78	3.71
13-24	451	.90	2.60	3.99
25-48	136	1.90	4.94	10.03
49-72	88	2.90	6.80	18.79
>72	215	12.71	20.05	47.00
Total/summary	11,655	.03	.85	.96

Table 3. Response times to wildfires. "Fires fought by the State of California" also were analyzed and found for the 6,278 fires, CAL FIRE had air tankers over the fires within one hour of the first report 96.7 percent of the time. This is in comparison to 37.9 percent of "fires not fought by the State of California." (Table created by Crowbar Research Insights LLC™ compiled with information from Fire Aviation, © Crowbar Research Insights LLC™. All rights reserved.)

The RAND Corporation analysis alone should drive policy makers to preposition aerial firefighting assets for rapid response to wildfire outbreaks. Policy makers should invest significantly more in aerial firefighting assets, especially SEATs and helicopters. They are the best and least expensive platforms available for success in an initial attack. Unfortunately, none of the RAND Corporation recommendations have been implemented by the FS.

Instead, as demonstrated below, since 2000, the FS intentionally reduced the number of LATs and VLATs utilized by federal firefighting agencies. Most experienced personnel involved with wildfire management would recognize that a current fleet of eighteen LATs and VLATs represents an insufficient number to fight more than 8,700 human and lightning wildfire incidents each year in the fifteen most fire-prone states. With so few aircraft, prepositioning the fleet to knock out the fire during the initial attack represents a pipe dream, as suggested by RAND's study. Moreover, the latest policy issued by the secretary of the USDA in January 2022 downplays the use of wildfire suppression or any additional investment in aerial firefighting assets (see figs. 2 and 3).

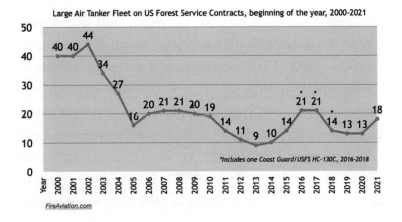

Figure 2. Number of FS large air tankers on exclusive use contracts at the beginning of each year. (Image by Fire Aviation, April 16, 2021, https://fireaviation.com/2021/04/16/what-did-we-learn-from-the-aerial-firefighting-use-and-effectiveness-study/.) [Photo credit: Wildfire Today © with copyright permission.]

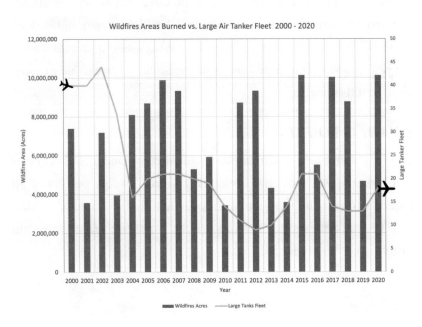

Figure 3. Wildfire areas burned versus large air tanker fleet, 2000–2020. (Figure created by Crowbar Research Insights LLC™ from data contained in the National Interagency Coordination Center annual reports, and FS budget authorization requests, © Crowbar Research Insights LLC™. All rights reserved.)

As figure 3 displays, forest fires consumed more than 10 million acres or 15,620 square miles (the equivalent of half of South Carolina) of forestland in 2020. The five-year destructive average is almost eight million acres per year. Three out of the past ten years (2011–2021), the United States experienced more than 10 million acres of wildland forest fire destruction (2015, 2017, and 2020).[5] Note the inverse correlation between the number of LATs and VLATs under contract, which declined from forty-four in 2002 to eighteen in 2021, and the increase in acres burned over that same period with a smaller air tanker fleet. To put these numbers in perspective, the estimated total loss of forestland over the past five years would equal the loss of almost 40 million acres, or 62,500 square miles (the total landmass of Georgia).

We found no evidence of any formal fleet plan to deploy enough aerial assets to provide more geographic cover necessary for the fifteen most fire-prone states, comprising 2.2 million square miles. That means the current fleet of eighteen air tankers would be required to cover 122,444 square miles each—a statistically impossible mission to protect this area.

UNABLE TO FILL

The two charts below summarize incident commander (IC) requests and FS responses for aerial firefighting assets. Twenty percent of all requests for air support went unfilled, identified as "unable to fill," or UTF. Unfortunately, 2019 was not the worst fire season in recent history (see figs. 4 and 5).

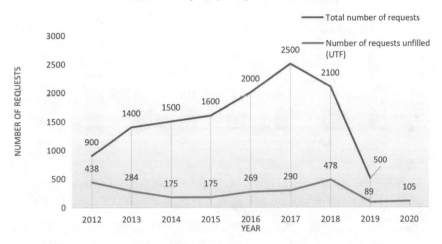

Figure 4. Unable to fill requests (UTF) summary. (Figure created by Crowbar Research Insights LLC™ from data retained by National Interagency Fire Center for large air tankers on exclusive use contracts, © Crowbar Research Insights LLC™. All rights reserved.)

Authors' Note

Figure 4 demonstrates a high demand for aerial assets by incident command teams through 2017. There was a sharp decline beginning in 2018 due to the high number of UTF requests that year (23 percent).

During the 2019 fire season, the incident command teams stopped contacting National Interagency Fire Center (NIFC) for aerial assets. In all likelihood, they realized it was doubtful aerial assets would be available. The scenario: a conditioned response of "why call, because the assets won't be available anyway."

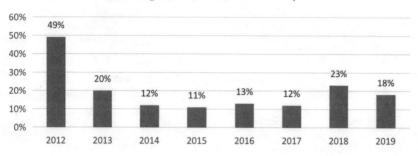

Figure 5. Percentage of unable to fill requests. (Figure created by Crowbar Research Insights LLC™, calculated with data from figure 4, © Crowbar Research Insights LLC™. All rights reserved.)

What private enterprise could exist if it fails to provide up to 25 percent of needed products to its customers? Certainly, successfully preparing an appropriate number of prestaged firefighting aircraft is anything but inexpensive. Still, failure to put out the fire during an initial attack is costly, as the RAND study proves. And as our analysis of mechanized equipment shows, despite the enormous taxpayer investment in predictive systems, the FS has failed to contract enough aircraft assets to put out the fires, particularly during the initial attacks.

Finally, we examined fleet operating practices. Current contracts call for only one crew per aircraft, with some latitude for a relief crew under certain conditions. As explained previously, mandatory pilot rest periods developed months in advance also constrict an aircraft's full use. To illustrate: During eight hours, LATs and VLATs typically complete two to five round-trip rotations (sorties) to a wildfire depending on the distance from the advanced base. Should operations commence at sunrise after the elapsed eight-hour flying period or following a fourteen-hour duty day, the crew must rest until the next day. This practice could leave the aircraft idle from midafternoon to sunset unless a backup crew is immediately available to continue operations. What if the contract stipulates a requirement for a second crew? In that case, we estimate an additional one to three sorties could be accomplished per day for each plane deployed, or a 40 to 60 percent increase in air tanker utilization.[6]

Despite the availability of avionics systems for certain current operating platforms to provide nighttime aerial firefighting, the FS prohibits nighttime deployment. While we acknowledge the safety concerns, nighttime aerial firefighting represents a future opportunity for a genuine next-generation air tanker platform. Currently, the Orange County, California, Fire Authority is undertaking nighttime helicopter retardant deployment to take advantage of lower wind and higher humidity conditions. This step represents a precedent to assess the possibility of success with nighttime suppression.

FLEET COMPOSITION

The FS fiscal 2022 aerial wildfire suppression fleet plan calls for the following essential use (EU) aircraft:

- 18 LAT (reduced from 44 in 2002—a 59 percent reduction)

- 60 SEATs (34 EU and 26 Amphibians)[7]

- 28 Type 1 helicopters (reduced from 34 in 2016—an 18 percent reduction)

- 34 Type 2 helicopters

- 46 Type 3 helicopters.

What capabilities differentiate the LAT aircraft with six platforms (and potentially eight) in the FS hangar—that is, B-747, DC-10, MD-87,

C-130, B-737, DHC-8/400, RJ-85, and BAe-146? We examined several benchmarks to analyze the available platforms, including

- the number of different make and model air tanker aircraft under FS contracts;

- age of the in-service aircraft (when was that type last manufactured); and

- cost to deliver water or retardant covering one mile of the fire line.

Modern aircraft fleet management practices focus on one or two platforms for overall management, maintenance, and operational efficiency. For instance, Southwest Airlines' use of only the Boeing 737 aircraft is an example of single platform. Unfortunately, as can be seen from our charts, the FS nominally uses six different, mostly aging, platforms.

We attempted to identify the FS's specifications to select and operate the most accurate, cost-effective, retardant delivery–capable aircraft. Despite contacting the NIFC directly and reaching out to individual aerial contractors, conducting exhaustive research, and examining the various requests for proposals, we only found a few specifications by airframe platform describing the FS's expectations. When such specifications are absent, how can aviation managers assure deployment of the best and most capable assets to meet the FS's mission to extinguish and control wildfires?

Definable and measurable metrics are essential for platform comparisons. While aircraft specifications allow selection of the best-performing platforms, they can be impractical from an operational standpoint. In the chart comparing effectiveness (included in the appendix), helicopter platforms outperform fixed-wing platforms; however, if the fire requires an extended transit, or there are no lakes to refill Bambi Buckets, the fixed-wing platforms become the logical airborne firefighting tool available to the IC.

To assess the current fleet, the following metrics were used to make comparisons of the FS aerial aircraft fleet (helicopters and fixed-winged aircraft):

1. amount of water and retardant lift capability (assuming no density altitude restrictions)

2. efficiency or success probability of the retardant drops by aircraft category

3. cost to drop water or retardant to cover a mile of the fire line

Here is a brief summary of our findings (our complete analysis is included in the appendix):

- By general aircraft category, helicopter drop efficiency is approximately 15 percent more effective than fixed-wing aircraft drops and particularly useful in an initial attack or in rescuing specific high-value elements, such as homes or critical business operations.

- The Boeing 747, though no longer immediately available, is the most successful and cost-effective VLAT fixed-wing platform to cover one mile of a fire line. Yet the FS terminated the B-747 contract, claiming inconsistent continuous flow of retardant, even though the independent operator later corrected the mechanical issue.

- In the current LAT and VLAT classifications, after the B-747, the DC-10 is the most cost-effective platform with a retardant-carrying capacity of 9,400 gallons.

- Of the current available fixed-wing fleet, SEAT air tractors represent the most cost-efficient and effective platform, particularly during an initial attack. But their low water and retardant capacities make them less effective should the wildfire become more extensive and increase in intensity.

- The third most financially viable platform is the B-737 (operated by Coulson Aviation).

- The BAe-146 (operated by Neptune) and the RJ85 (operated by Aero Flite) round out the current top four.

- The Coulson Aviation C-130 and the Aero Air MD-87 fall into last place.

We found little to no evidence that federal firefighting contracting officers establish these specifications to engage the best fleet from a cost-per-mile and retardant-dropping efficiency perspective. The only specification named in the contracts we reviewed relates to the consistency of the retardant flow rate during a drop.

> **Authors' Note**
> As outlined in the appendix, our work indicates the cost to drop retardant to develop one mile of fire line can range from $19,167 for the SEAT Air Tractor to $146,740 for the MD-87. Such a significant cost difference merits changes to contract specifications and commitments for the country's aerial firefighting aircraft fleet. Instead, the same independent operators are awarded EU and call-when-needed (CWN) contracts each year, apparently without regard to cost per mile, efficiency requirements, or other generally understood industry qualifications.

Other necessary specifications include retardant-carrying capacity, drop accuracy, tight turns, maneuverability, proximity to ground level, performance during turbulence, loiter (stay aloft) time, and gravity (g) force ratings on aircraft wings at different weight and stress limits. It is difficult to ascertain if the FS considers these critical factors when awarding aircraft contracts. These criteria are not mentioned in the *FS Aerial Firefighting Use and Effectiveness Report*. Absent specific evaluation of these factors, safety concerns escalate significantly. Aircraft cycles—that is, takeoffs and landings—can take their toll on airframes. With the exception of the C-130s, the current fleet is comprised of former commercial aircraft converted into aerial firefighters, meaning stress levels to airframes and wings take on heightened scrutiny due to increased safety risks. The frequency of wing replacements, engine overhauls, and aircraft frame maintenance to comply with manufacturing specifications of a commercial airliner differ entirely from those required to maintain a firefighting air tanker subjected to potentially higher load and stress factors.

Finally, the lift capability of the platform determines how many sorties the platform must accomplish to drop water or retardant on a mile-long fire line. The return and replenishment times become critical should winds expand the flames. Time is vital.

CAL FIRE

Though this book is focused primarily on FS asset management, acquisition, and utilization, several states have a capable and effective state-owned airborne firefighting forestry service. California is one of them. This state seems to have recognized a potential solution to operating aerial assets, unlike the FS. California maintains the largest fleet of firefighting assets of any state through its Department of Forestry and Fire Protection, called CAL FIRE. Its fleet of owned aerial firefighting assets includes

- twenty-three Grumman S2-T 1200-gallon air tankers,

- twelve UH-1H super Huey helicopters, and

- fourteen OV-10A air tactical aircraft.[8]

In 2021 CAL FIRE announced the addition of twelve Sikorsky S70i Fire Hawk helicopters, which, along with seven Lockheed Martin C-130H Hercules air tankers scheduled for delivery in 2022 and 2023, are examples of CAL FIRE's commitment to invest in its firefighting aviation program. CAL FIRE also contracts with the same aerial firefighting companies as the FS for LATs and VLATs assets under EU and CWN contracts.[9]

Unlike federal contracts, CAL FIRE executes unique third-party agreements to furnish the flight crews and provide procurement, maintenance, and logistics support for its fleets.

Except for CAL FIRE helicopters, which are flown by its employed pilots, CAL FIRE's support contractor DynCorp International/Amentum provides air tanker, air tactical fixed-wing pilot services, and all aircraft maintenance services for CAL FIRE–owned assets. The current DynCorp contract covers three years, with a two-year extension option through 2025. The DynCorp contract value is $352 million, or approximately $70 million per year.[10] LSI Logistic Specialist, Inc., provides procurement and parts management services for the entire CAL FIRE fleet. Although unpublished, we estimate the LSI Logistic Specialist contract to have a value of about $45 million through 2023, around $15 million per year. Based on these contract values, the current CAL FIRE–owned fleet of thirty-seven fixed-winged aircraft incurs an average cost per plane for crews, maintenance, and support of $2.3 million per aircraft.

All CAL FIRE aircraft are stationed at fourteen air attack and ten helitack bases located throughout the state. CAL FIRE projects that air assets can reach most wildfire locations within twenty minutes. As stated in our analysis of preflight staging to launch the national fleet, CAL FIRE's twenty-minute time line fails to account for the time to prep the plane, recharge it with water or retardant, or complete its preflight safety checks. Compared with the nimbler Grumman S-2s, a more likely estimate is thirty to sixty minutes of prep time and another twenty to twenty-five minutes to reach most wildfire incidents—a total cycle time of fifty to ninety minutes.

ASSESSMENT AND RECOMMENDED FS FIXED-WINGED FLEET SIZE

A clear mission statement is required to make an accurate determination of fleet composition and size. Unless the statement and related interagency memoranda of understanding are changed to call for a "put out the wildfire" strategy, the current FS fleet size, composition, and suppression practices will continue. As a result, we can expect wildfire seasons to increase with larger, more intense, and even smokier incidents. On the other hand, an aggressive wildfire suppression mission calling for massive "warlike" actions against wildfire outbreaks would change composition of the aerial firefighting fleet, the number of dedicated aircraft, crews, maintenance, and fleet support.

The American public should be shocked that this country engages only eighteen LATs to cover 3.8 million square miles of landmass. By contrast, Israel engages fourteen SEATs to cover its landmass of only 8,550 square miles (the same size as New Jersey).[11] Unfortunately, the FS wildfire policy refuses to acknowledge the critical importance of aerial firefighting assets, thereby violating its founding principles: to protect and preserve our national forests. From our research and discussions, it is evident the principal focus is on ground personnel. While we acknowledge that ground crews put out the wildfires, the situation only will worsen due to poor utilization of both mechanized equipment and the deployment of the appropriate aerial air tankers.

We strongly support a massive "warlike" response and mission statement. Under this scenario, we used our proprietary model to assess SEATs and LATs as the primary aerial assets for an initial attack wildfire incident response. SEATs bring to this "war" excellent drop accuracy, low capital and operating costs, maneuverability, quick response, and water and retardant capabilities. LATs give us expanded availability of larger retardant deployment capacities. Until a new next-generation platform is brought to market, we acknowledge that current LATs are the best, but not optimal, solution.

Our model assumes an immediate deployment of three SEAT aircraft and a ready deployment of up to six SEATs from other advanced air bases to extinguish small fires before they become wildfires. Water-filled SEATs will deploy immediately and simultaneously with firefighting crews when the 911 call is activated at the local fire station. Dispatchers and the IC would determine the actual number of aircraft deployed, understanding that SEAT deployment is inexpensive insurance and, working with the initial attack ground firefighters, the best opportunity to quickly extinguish the blaze. To mobilize three water-filled SEATs to an incident 100 miles away and complete a full return-trip rotation to the advanced base, we estimate the cost to be approximately $57,500. This estimate assumes ownership of the air tractor and contracting for the crews, maintenance, and support. Swarming the wildfire with two additional squadrons of three SEATs from other nearby advanced bases is also a scenario built into our model. A squadron of nine SEATs should cost less than $250,000 per incident. The RAND study referenced earlier projects "a successful initial attack can save $4.3 million per incident."[12] The cost-benefit analysis of deploying a squadron of nine SEATs defies any logical counterargument.

Ground crews and SEATs cannot always contain or extinguish wildfires, especially during the initial attack. Our model assumes LAT and VLAT assets would be deployed in a continuous stream, along with a squadron of LATs and VLATs from other advanced bases to assist. The goal is to establish a constant fire line perimeter of retardant around the wildfire more quickly than what occurs today.

For SEATs, we used estimated rotation and cycle times, retardant tank capacities, cost-per-mile data, and historical wildfire incident data to determine the optimum fleet size for the fifteen most fire-prone states. We believe optimal configuration is a fleet of three SEATs attached to each of the 33 advanced bases in the fifteen states and one to two SEATs attached to each of the other 112 advanced bases.[13] Conservatively, we need 200 SEAT aircraft nationwide available to attack early discovery wildfires. There currently are 34 wheeled and 26 amphibious SEATs under contract with various federal agencies, which means 140 air tankers should be acquired to increase the SEAT fleet.

For LATs, their expected rotation times, cost-per-mile operating costs, and drop accuracy form the metrics for our scenarios, again focusing on wildfire incidents in the fifteen most fire-prone states.

Based on our analysis, we believe deployment requires a total fleet of 75 to 125 LATs, three each at the 33 advanced bases and the rest assigned to the remaining advanced bases. Our model assumes LAT tank capacities of 4,000-gallon platforms.

Finally, for VLATs, our model suggests a fleet size of 30 to 50 planes at the 33 advanced bases. These assets can rotate to other bases depending upon fire conditions.

To become more efficient, the FS needs to own the entire aerial firefighting fleet and adopt the CAL FIRE business model. At the same time, crews, maintenance, procurement, and support services must be acquired through three- to five-year contracts with third-party providers. A business framework that adopts the CAL FIRE model is more effective, costs less to the taxpayers, and provides better fleet management.

The table below summarizes our proposed fleet plan, cost to acquire, and annual operating costs (see table 4).

Aircraft Type	Total Number	Acquisition Cost (in Millions)	Crews and Maintenance Annual Cost (in Millions)	Total Annual Cost Including Depreciation(e) (in Millions)
For Each Aircraft				
SEATs	1	$2.0(a)	$2.5 M(b)	$2.6
LATs	1	$7.0(c)	$1.6 M(d)	$1.8
VLATs	1	$24.4(c)	$3.0(d)	$3.8

Aircraft Type	Total Number	Acquisition Cost (in Millions)	Crews and Maintenance Annual Cost (in Millions)	Total Annual Cost Including Depreciation (in Millions)
Total Fleet				
SEATs	200	$400.0	$500.0	$547.6
LATs	100	$700.0	$160.0	$180.0
VLATs	40	$976.0	$120.0	$151.5
	Totals	$2,076.0	$780.0	$879.1

Table 4. Proposed fleet plan. (Table created by © Crowbar Research Insights LLC™. All rights reserved.)

a. Current acquisition cost of an Air Tractor 802A, fitted with 800-gallon tanks per manufacturer—Air Tractor, Inc.

b. Used CAL FIRE's contract rate with DynCorp and LSI Logistics, adjusted for inflation.

c. For LATs used the estimated current market price of B-737 for airframe, engines, avionics, and 4,000-gallon retardant drop system. For VLATs used an estimated current market value price of a 747-400 for airframe, engines, avionics, and a 19,200-gallon retardant drop system. It also assumes any refurbishment costs to put into service. Source of airframe and operating costs data-Avitas, Inc., Jet Aircraft Values, March 2020.

d. Based on block hour costs reported by airline service providers for B-737-700 and B747-400, excluding depreciation. It also includes an inflation adjustment and a profit margin for the contractors providing the outsourced services Source: Avitas, Inc., Aircraft Block Hour Operating Costs and Operations, Second Quarter 2020.

e. We assumed a 25-year depreciable life and 15 percent residual value, the same depreciation period and residual value accounting policy adopted by Southwest Airlines. The total annual costs noted above include $0.100 million for SEATs, $0.238 million for LATs, and $0.830 million for VLATs related to these depreciation costs.

All information was computed using the same 100-day MAP contract periods for SEATs and 160-day MAP contract periods for LATs and VLATs for comparison purposes to the existing contracts with independent operators.

Every eight years, one-third of the fleet should be replaced with newer aircraft, including the next-generation air tanker. For example, the *FS Aviation 2020 Annual Report* sets forth costs paid to contractors for the 13 LATs. The total cost in 2020 for the LAT and VLAT fleet amounted to $206,590,262 or $15,891,560 per air tanker. This cost includes the aircraft, crew, maintenance, and support paid to the independent operators.[14]

Comparing this information to the cost of ownership in table 4, for LATs, taxpayers would save, on average, more than 85 percent of the funds the FS currently spends for LATs—that is, $15,891,560 versus $1,800,000. This cost-saving analysis is similar in both 2020 and 2021. Based on a fleet of 18 LATs and VLATs, total savings exceed $250 million per year.

Consider this: at $250 million annual savings, the payback for the entire proposed fleet of 140 LATs and VLATs totals eight years. Further, prepositioning the SEATs, LATs, and VLATs based upon highly predictable lightning-caused wildfires in the fifteen states should reduce both number and 300-acre size of wildfire incidents dramatically. We desperately need to drastically reduce the $300 billion annually lost from wildfires. In our view, the investment of $2.1 billion in the proposed fleet makes the cost-benefit analysis compelling.

Nevertheless, we recognize wildfire management also requires investment in well-trained IC teams, fire crews on the ground, and timely deployment of mechanized equipment to actively suppress wildfires. Further, we acknowledge the parallel role of investing in improved forest management practices so necessary for reducing wildfire risk. In chapter 11, we propose a private-public partnership model to finance each of these critically important initiatives.

THE NEXT-GENERATION AIR TANKER

Although far from optimal, converted military and commercial aircraft represent today's only viable air tanker options until a truly next-generation air tanker is developed. Using these converted aircraft does not define a truly "warlike" strategy.

For more than seventy years, the FS engaged aerial air tankers to suppress wildfires. During that entire period, however, FS personnel have failed to establish or propose development of a true aerial air tanker platform specification, present those specifications to the aircraft-manufacturing community, and create a bidding process to establish such a platform. Now is the opportune time to act.

The financial analysis provided in this chapter highlights the lack of initiative to create the next-generation aerial air tanker-firefighter. A worldwide need is apparent and demand would favor economies of scale, which could reduce the cost per air tanker by the selected manufacturer.

Several air tanker platforms under development or recently released are listed in the appendix. The Embraer KC-390 may have promise, but its 4,000-gallon retardant capacity translates into an expensive cost-per-mile metric. Converting passenger Boeing-757s or Airbus-A320s developed more than thirty years ago does not appear to be the best long-term air tanker strategy. According to the *Encyclopedia Britannica*, the Boeing 757 was introduced into service on January 1, 1983, and the Airbus A320 in 1988.[15]

Our models suggest a minimum of three retardant-drop capacity platforms: the first—5,000-gallon retardant-drop capacity; the second—10,000-gallons; the third—20,000-gallons. With these capacities, aircraft usage can successfully cover the variety of terrains and Wildland Urban Interface (WUI) conditions in the United States while significantly reducing the cost curve per mile.

We view the next-generation air tanker as a considerable opportunity to develop the needed specifications based on metrics of speed, drop accuracy, proximity to ground level, maneuverability, fuselage, and wing g-force strength, tight turns, and cost per mile. FS leadership should play a pivotal role in engaging the global firefighting agency community to develop the next-generation air tanker platform for an aerial firefighter.

THE KEY TAKEAWAYS FROM THIS CHAPTER ARE:

- *Put out the fire.* There can be no compromise on the mission of the firefighting services of the FS. The perilous attitude about containing a fire has cost the United States millions of acres, countless lives, and billions of dollars in annual damage. Leadership must change its attitude about extinguishing wildfires first by making it a "war effort." Extinguishing wildfires early can save over $20 billion in damages yearly.

- Helicopters and SEATs should be deployed simultaneously to any wildland fire report along with local firetrucks. The key to controlling and quickly extinguishing a wildland fire is immediate application of water—sometimes a great deal of water. A small fire can be extinguished quickly with the right mix and number of helicopters and SEATs.

- Prepositioning the correct number of SEATs and air tankers through substantial investment in existing predictive systems requires immediate attention. Allowing this combination of lack of knowledge, poor planning, and inadequate size of aerial firefighting assets is inexcusable.

- The current level of available firefighting assets (SEATs, LATs, and VLATs) is inadequate to meet the escalating number of wildfires. There should be 200 SEATs aircraft available, at a minimum, which is triple the current number of SEATS deployed.

- Speed, number of SEATs deployed, and proximity to the fire are critical to a successful extinguishing effort during the initial attack.

- Expand the number of LATs and VLATs from the current inadequate fleet of 18 to a minimum of 140 aircraft.

- Current contracting methodology should be scrapped for a CAL FIRE aircraft ownership model. Government agencies should purchase aerial assets and contract for the crew, operations, and maintenance functions. This practice could save taxpayers above $250 million per year, representing an 85 percent annual cost savings.

- Address the current pilot shortage. The FS is not at the top of the list when it comes to hiring well-qualified career pilots. Where is the pilot support mechanism by the FS and supporting contracted companies that will ensure an adequate supply of pilots given the rapidly increasing number of wildfires?

- The FS must take a global leadership role to develop a next-generation air tanker platform designed explicitly for worldwide wildfire use. Examine optional retardant capacity platforms that take into account WUI and geographical characteristics.

- Invest in an expanded aerial air tanker fleet to provide a substantial cost-versus-benefit ratio with the potential to reduce the $300 billion each year in wildfire damages.

CHAPTER 8

INTERNATIONAL WILDFIRE AND FORESTLAND MANAGEMENT OBSERVATIONS

The 2020 global wildfire season brought extreme fire activity to the western U.S., Australia, the Arctic, and Brazil, making it the fifth most expensive year for wildfire losses on record.[1]

Wildfires are not just a national issue; they are an international problem affecting every region across the globe. Wildfires ravage many diverse areas, such as the Amazon, Mediterranean, Siberia (Russia), China, Southeast Asia, Australia, and the United States. While this book focuses on the prevention and management of U.S. wildfires, its information presents a graphic picture of the destructive impact of wildfires worldwide. It also compares specific international practices with current U.S. forest management methods.

The 2020 Worldwide Fund for Nature/Boston Consulting Group study offers this global analysis: "85% of the world's surface area burned each year is in tropical savannahs which make up 19% of the total land cover of the globe. In tropical and subtropical regions, forest fires are mostly intentionally set for land-use change clearing and preparing new areas for cultivation. Farming, using slash and burn methods occurs in many countries, especially in Southeast Asia and Africa, where trees are cut and burned to expand arable land while enriching its soil with ashes in nutrients. Prescribed burns are also used on larger-scale agriculture projects, clearing Indonesian acreage for palm oil plantations and cattle ranches in Brazil—and these fires frequently run out of control, leaving soil erosion risk in their wake. In the Brazilian Amazon, fires are also part of a pattern of increasing encroachment into public and Indigenous Peoples' lands."[2]

The table below summarizes the substantial human and economic losses from global wildfire disasters, in cumulative totals for the years reported (see table 1).

Region	Wildfire Deaths	Estimated Damages (U.S. $ Billions)
Africa	233	$ 4.1
Australia	255	7.8
Canada	316	18.0
China/East Asia	177	0.5
Europe	330	11.3
Indonesia	67	24.0
Middle East	28	3.3
Russia	464	10.5
South America	243	8.5
United States	724	66.9

Table 1. Human and estimated direct losses from global wildfire disasters, 2000–2021 (twenty-two years), except for the United States, with data only from 2015–2021 (seven years). (Table by Centre for Research on the Epidemiology of Disasters, EM-DAT: The OFDA/CRED International Disaster Database, https://www.emdat.be/cred-crunch-55-volcanic-activity-wildfires.) [Table credit: Reports from CRED supplemented for additional public press releases of disasters not reported by CRED, by Crowbar Research Insights, LLC. Estimated Damages are estimates of incremental costs and reported in the CRED data base adjusted for recent years based on press reports.]

> **Authors' Note**
>
> The information in table 1 requires explanation. The database we accessed did not report many wildfire events after 2015 for countries other than the United States. To make the analysis meaningful, we added deaths and estimated damage data that occurred after 2015 for every country that had data available from other research.
>
> The United States data surprised us by reporting more deaths and damages in the last seven years than any other country or continent during a period of twenty-two years (a time period three times longer than the United States). Compared to all other countries and regions, the United States ranks the most destructive in terms of deaths and economic loss.

ASIA/INDONESIA

The Great Black Dragon Fire of 1987 is considered one of the largest and deadliest wildfires in the world. It killed 200 people and destroyed more than 3 million acres in China while scorching 18 million acres in Siberia.[3] Indonesia, the Southeast Asian country most affected by wildfires, reported 2.12 million acres destroyed in 2019, the most since 6.4 million acres were consumed in 2015.[4, 5] The World Bank estimated the cost to Indonesia at $5.2 billion (U.S.). The 2015 fires cost a staggering $16.1 billion (U.S.) in direct and indirect costs.[6] Since 1990, Indonesia, a country that once primarily consisted of forestland, has lost 11.1 million acres of forests due to logging, deforestation, and wildfires.[7] Elsewhere, a 2019 wildfire in the Gangwon Province of South Korea destroyed more than 1,880 buildings, with damages estimated at $4.6 million (U.S.).[8] Thailand experienced a 17 percent increase in land burned by wildfires in 2020 as compared to 2019, with wildfires consuming 2.6 million acres in 2020.[9] Wildfires in 2020 also affected Myanmar, Cambodia, Laos, and Vietnam.[10]

On the Indonesian islands of Sumatra and Kalimantan, a recent scientific study estimated that only 10 percent of the tropical forests remain resistant to wildfire risk. Humid and cool microclimates maintained by these tropical forests act as fire barriers. Due to severe forest fragmentation caused by wildfires covering the size of the Netherlands from 2014 to 2019, the remaining forests no longer maintain the microclimates needed to resist wildfires naturally.[11]

AUSTRALIA

A 2016 analysis discovered a 40 percent increase in Australian bushfires since 2011.[12] As bad as this was, the increase paled compared with the horrific Black Summer bushfires that burned a staggering 46 million acres in Australia from June 2019 to March 2020, almost five times the acres destroyed in the United States in 2020. The fires affected every state and territory in Australia, killed 34 people, and destroyed more than 10,000 structures, including 2,779 homes.[13, 14] Nearly 3 billion animals were displaced or killed, including wallabies, koalas, and kangaroos (see fig. 1).[15]

Figure 1. A lone wallaby foraging in a burned forest outside Mallacoota, Australia. (Photograph by Jo-Anne McArthur via We Animals Media, February 14, 2020, https://weanimalsmedia.org/2020/02/14/disaster-response-how-australia-is-failing-its-wildlife/.)

Air quality dropped to hazardous levels in many parts of Australia and contributed to an estimated 3,500 deaths and hospitalizations.[16] Smoke from the bushfires traveled as far as South America and possibly the Antarctic.[17] These fires emitted more than half of Australia's annual carbon dioxide emissions.[18] Of particular significance, the bushfires occurred in places where they do not commonly occur, such as in the rain forest and banana plantations as well as in areas prone to bushfires (see fig. 2).[19]

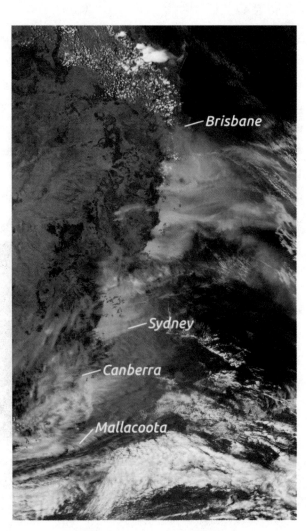

Figure 2. Imagery from NASA Aqua Satellite showing MODIS and VIIRS data of Australian bushfires on December 7, 2019. Highlighted in red are fire detections notable landmarks labeled. (Photograph by NASA Worldview application, part of the NASA/EOSDIS, December 7, 2019, https://commons.wikimedia.org/wiki/File:2019-12-07_East_Australian_Fires_Aqua_MODIS-VIIRS-LABELS.png.)

The fires caused $3.5 billion (U.S.) in damages and business interruptions, making it the costliest wildfire season in Australia's history. The 2009 wildfire season, known as the Black Saturday Bushfires, is the second most expensive, with $1.2 billion (U.S.) in damages and more than 1.1 million acres burned.[20]

The world's largest volunteer firefighting organization is in New South Wales, Australia. The NSW Rural Fire Service has worked with local agencies to prevent and fight bushfires for more than one hundred years.[21] Committed to an aerial firefighting strategy, Australia's National Aerial Firefighting Centre (NAFC) has a dedicated fleet of 160 aircraft and additional 350 aircraft available. Australia's fleet consists primarily of single-engine air tankers (SEAT) aircraft and helicopters, along with fire trucks and trained volunteer and career personnel.[22] Even with these resources, reservists from the Australian army, navy, and air force, and firefighters and equipment from New Zealand, Canada, Singapore, and the United States were called upon to fight the 2019–2020 bushfires.[23] Due to the severe need for air tankers, in 2020, the Australian government entered into expensive $20 million (U.S.) short-term leases with a U.S.-based aerial firefighting company to assist in combating the bushfire disaster. Subsequently, Australia purchased one Boeing 737 air tanker as a permanent addition to its fleet; nonetheless, the aircraft acquired fails to provide the best solution for an aerial firefighting mission.[24] In 2021 NAFC released its *National Aerial Firefighting Strategy, 2021–26*, acknowledging the need for additions to its large air tanker fleet, along with analysis of new aircraft platforms that may become cost-effective options, including ex-military aircraft.[25]

EUROPE

Historically, the Mediterranean area of Europe experiences the most wildfire activity, while northern countries—Finland, Germany, Switzerland, Norway, and Sweden—have the fewest wildfires (see table 2).

Europe's Wildfires from 1980 to 2016 (Most and Least by Country)		
Country	Per Year	Acreage Destroyed Each Year
Most Incidents		
Spain	10,814	400,119
Portugal	7,127	265,382
Italy	7,040	260,462
Greece	3,000	110,985
France	1,647	60,929
Least Incidents		
Finland	64	1,349
Germany	66	1,719
Switzerland	110	925
Norway	143	2,282
Sweden	305	6,708

Table 2. Summary of Europe's wildfires from 1980 to 2016. (Table created by Crowbar Research Insights LLC™ with information extracted from https://effis.jrc.ec.europa.eu/applications/data-and-services, © Crowbar Research Insights LLC™. All rights reserved.)

The annual average burned area in the Mediterranean region has quadrupled since the 1960s.[26] Spain and Portugal experience the most wildfires in this region, and its peak wildfire season now extends an average of two months longer than in previous years to five months per year.[27]

In 2003 wildfires destroyed more than 10 percent of Portugal's forests. The deadliest Portuguese season was 2017, when 114 people perished in two separate wildfires.[28, 29] Spain experienced its most extensive wildfires in twenty years in 2019.[30] In Greece, a 2018 wildfire killed 104 people and damaged or destroyed 1,650 homes, and during 2021, more than 250,000 acres were destroyed. The extensive destruction is partly because of wildfires in wildland-urban interface (WUI) areas.[31]

Wildfire management practices vary in these countries, with most using firefighting equipment, tanker planes, and helicopters for fire suppression. Spain, Portugal, Italy, Greece, and France all use SEATs, helicopters, and water-scooper aircraft as a significant component of their wildfire suppression strategies.[32] Cooperative agreements between many Mediterranean countries, including Spain and Portugal, authorize sharing of wildfire firefighting assets during peak disaster incidents. In 2019 Portugal announced plans to acquire several Brazilian made Embraer C390s—perhaps the most advanced aerial firefighting aircraft—to deploy retardant on its active wildfires.[33] Additionally, many countries in this region use prescribed burning as a fire management approach; however, this practice is not allowed in Greece.[34]

In addition to wildfire suppression, active forest management in several European countries keeps forests resilient from wildfire risk. For example, due to severe drought, beetle infestation, and a large wildfire in 2018, Germany determined through satellite monitoring that about a third of its forests were in perilous condition. Federal and state funding of $878 million allowed replanting of more than 440,000 acres of trees. Because a quarter of German forests consist of spruce trees, typically used for timber harvesting, the replanting plan incorporated mixed species to make the forests environmentally more resilient. Forestry officials acknowledge they are using this time to determine the "right mix" of trees but insist decisive action now provides the basis for continuous improvement to future forest management practice.[35]

RUSSIA/SIBERIA

The most extensive forest cover on Earth is in Russia. Russia's forestry management practices include clear-cutting, leaving their forests more susceptible to wildfires. Almost 50 percent of timber is logged illegally in some Russian forests.[36] These forests are primarily federal property and are managed by the Federal Agency for Forestry (Rosleskhoz), a part of the Ministry of Natural Resources and Environment.[37] The 2003 Siberian Taiga fires burned 47 million acres.[38] Wildfires near Moscow in 2010 damaged or destroyed 2,500 homes and killed at least 50 people (see fig. 3).[39]

Figure 3. CBS News documented forest fires in the Krasnoyarsk region, Siberia. It is unambiguous evidence of a forest management emergency: the northern landscape transformed by heat and fire. (Used with permission.)

Record heat in Siberia contributed to record wildfire seasons in 2020 and 2021, which saw the highest incidence of activity in the June-October period since 2003. Fire officials estimated that the 2020 wildfires burned an area larger than the size of Greece and released 35 percent more carbon dioxide than emitted in the entire previous year.[40, 41] 2020 also marked the worst wildfire season to date in the Chernobyl region of Ukraine.[42] In 2021 Greenpeace reported that Siberian wildfires in the Yakutia forests destroyed 3.7 million acres.[43]

SOUTH AMERICA

Most of the Amazon is in Brazil, where recent wildfires were the worst in a decade.[44] July 2020 saw a 55 percent increase in wildfires as compared to the average number of wildfires in July 2010 through 2019. All of this occurred just one year after 2019 set record numbers, with more than 72,000 wildfires, an 83 percent increase from 2018 and the highest since Brazil started keeping records in 2013.[45, 46] The majority of these wildfires are linked to deforestation practices occurring in this region (see fig. 4).[47] In the Amazon regions of Colombia, Peru, and Venezuela, the number of wildfires in 2020 was higher than in 2019.[48] Rampant wildfires also affected Argentina, Bolivia, Brazil, and Paraguay in 2020.[49] Community management of forests makes enacting national policy difficult, particularly in countries such as Brazil.

Figure 4. Fires burning in the Amazon on August 17, 2020, next to the borders of the Kaxarari Indigenous territory, in Labrea, Amazonas state. Felled forests are intentionally lit in the Amazon to clear land for cattle ranching. (Used with permission.)

The figure below summarizes the number of wildfires recorded by each South American country in 2021 (see fig. 5). South America's nations rely almost entirely on localized decision-making to deploy firefighting resources. Responses include soldiers and some firefighting planes capable of deploying suppression retardants.[50, 51]

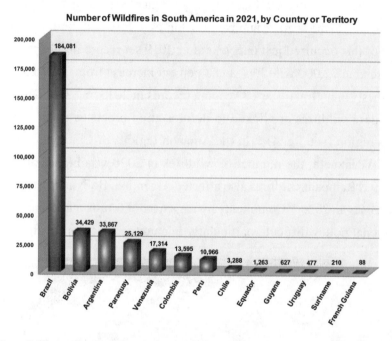

Figure 5. The number of wildfires in South America in 2021, by country or territory. (Figure by Crowbar Research Insights LLC™ using data from , https://queimadas.dgi.inpe.br/queimadas/portal-static/estatisticas_paises/, © Crowbar Research Insights LLC™. All rights reserved.)

Unfortunately, politics play an impactful role in forest management practices. For example, over 35 million acres of the Amazon forest could lose their protected status due to Brazil's 1965 Forest Code changes. Changes in this law in 2012, approved by the country's Supreme Court in 2018, allow rural landowners the right to clear large areas of privately owned native vegetation previously slated for conservation. The 1965 Forest Code established an international best practice by setting aside between 20 and 80 percent of native forests and savannas as conservation units and Indigenous reserves in specific country regions. Large agribusiness, illegal loggers, and mining interests stand to gain. At the same time, Amazon forests and the environment lose thanks to government officials who appear unwilling to curtail the environmental damage worsened by political agendas despite local business and international pressure to stop deforestation of the Amazon created by the 2012 legislation.[52] Moreover, international efforts to curtail Amazon deforestation through a $1 billion (U.S.) multicounty Amazon fund face challenges due to political posturing. Only two of the original partners (Germany and Norway) nowparticipate.[53]

AFRICA

A 2020 United Nations (UN) forest assessment report estimates Africa had the world's most significant rate of deforestation from 2010 to 2020 at 9.6 million acres per year, followed by South America at 6.4 million acres annually.[54] Africa is sometimes referred to as the "fire continent." Its tropical savannah landscape contains accumulated undergrowth from the country's rainy season that becomes highly flammable during the dry season. The fire season in southern Africa runs from August through November and from December through March in central Africa. NASA has estimated that 70 percent of worldwide fires occur in Africa, but the primary source is not lightning.[55] Most African fires are due to agriculture and land clearing, especially in savannahs. Savannahs (grassland areas) cover about half of Africa's surface. While wildfires are common in savannahs, ironically Africa is experiencing a decline possibly due to the increased use of savannahs for agricultural purposes.[56]

In August 2019, two African countries—Angola (6,902 wildfires) and the Democratic Republic of Congo (3,095 wildfires)—reported more wildfires than in Brazil (2,177 wildfires). There was widespread international media coverage regarding the high number of African wildfires; however, according to officials in these countries, the fires were "small slash-and-burn operations" and not forest fires. Additionally, fires are part of the southern African ecological system and occur annually in the savannah. In Mozambique, about 700,000 acres of forests are burned each year, primarily due to intentional slash-and-burn operations. These prepare the land for crop planting but strip the soil of nutrients and lead to destructive erosion. The situation is alarming in Madagascar, where only about 10 percent of its forests remain due to slash-and-burn farming that destroys almost 300,000 acres annually. Madagascar's Environment Minister Alexandre Georget predicted Madagascar "will be completely deforested in 40 years."[57]

More recently, in April 2021, a wildfire in Cape Town, South Africa, destroyed a university library of African antiquities and the oldest working windmill in this country. According to a statement from national park operator SANParks, "One of the major contributors to the rapid rate of spread was the ancient pine trees and their debris. The fire created its own wind that further increased the rate of speed."[58]

There is sparse information about African forest and wildfire management practices, limiting the opportunity to analyze the scope of the problem and develop action plans to reduce the incident risk. Some techniques exist while others are evolving. In Kenya and Tanzania, fires set by Maasai pastoralists promote green growth for grazing animals.[59] In the Congo and Guianas regions, the World Wildlife Fund (WWF), working with a technology consortium, expects to deploy an early warning system to predict and prevent illegal deforestation.[60] Additional critical issues in managing wildfires include the lack of national or continent-wide wildfire management policies, designated governmental agencies to manage or aid with wildfires, and the absence of cooperative wildfire agreements with the various jurisdictions throughout Africa.[61]

CANADA

Canada's boreal forest is the world's largest intact forest ecosystem. It stretches across 1.2 billion acres of northern Canada from the Yukon to Newfoundland and Labrador. Moreover, it represents 25 percent of the world's remaining intact forest, even more than the Amazon rain forest (see fig. 6).

Figure 6. North American's Boreal Forest. (Image from the Pew Charitable Trusts, March 19, 2015, https://www.pewtrusts.org/en/research-and-analysis/articles/2015/03/19/fast-facts-canadas-boreal-forest.) [Image credit: Copyright © 2015 The Pew Charitable Trusts. All Rights Reserved. Reproduced with permission. Any use without the express written consent of The Pew Charitable Trusts is prohibited.]

An estimated 8,000 wildfires burn 1 million acres each year in Canada.[62] The 2014 wildfires burned twice as many acres compared to the previous decade's average.[63] The Northwest Territories fire of 2014 burned 1.4 million acres of forestland, the most in a single fire season, and led to poor air quality across the country and even in a part of the United States, with smoke traveling as far south as North Dakota.[64,65] British Columbia experienced record wildfires in 2017 burning 500,000 acres. Fighting these fires cost $649 million (Canadian). The 2018 wildfires surpassed 2017 figures, burning 550,000 acres, damaging 2,211 properties, and costing $615 million (Canadian) in fire suppression.[66]

In only a few years, the increased severity of wildfires negatively impacted local ecosystems. These wildfires removed the ground layer of residential organic matter (biome), allowing jack pines to proliferate, dominating portions of the forest and thereby harming other vegetation and diminishing biodiversity.[67]

Canada actively uses aerial fighting assets as part of its wildfire suppression strategies, particularly in more fire-prone provinces. Manitoba owns a fleet of aerial firefighting assets and arranges with individual contractors to provide crews and maintenance covering ten-year periods. This process represents a best practice in our view because of the capital-intensive nature of acquiring expensive, specially designed aircraft.

The Canadian government determined in 2020 that its wildland fire management agencies spent between $800 million to $1.4 billion (Canadian) annually for wildland fire protection during the past decade. For six of the last ten years, the cost of wildland fire protection exceeded $1 billion (Canadian). Not surprisingly, the report concluded that actively suppressed wildfires in forested areas were not as high compared to the high wildfire activity in inhabited (WUI) forest areas.[68] Government forecasts expect Canadian suppression costs to increase by $150 million (Canadian) per decade due to WUI, particularly in western Canada.

GLOBAL FOREST MANAGEMENT AND WILDFIRE SUPPRESSION BUDGETS

The table below summarizes the latest information about individual country commitments to fund forest management and wildfire suppression efforts (see table 3). On every continent, financial issues hinder effective management as forestland management programs are slashed or eliminated.[69] U.S. information is excluded here because chapter 2 includes a detailed analysis.

Country Forestry Budget vs. Total Country Budget Expenditures (Excluding U.S.)			
Country	Country Budget (in Billion U.S. $)	Forestry Budget (in 2021 Billion U.S. $)	Percent (%)
Australia	441.7	5.8	1.31
Canada	411.5	2.03	0.49
China	552.9	15.38	2.78
Europe			
Spain	724.6	2.76	0.38
Portugal	122.0	0.123	0.10
Italy	1,214.8	2.37	0.20
Greece	124.0		N/A
Finland	80.6	3.2	3.97
Germany	490.2	1.00	0.20
Indonesia	540.9	0.500	0.09
Russia	294.5	1.31	0.44
South America			
Brazil	328.3	0.79	0.24
Chile	40.1	0.09	0.22
Venezuela	0.6	0.0001	0.01

Table 3. Individual country budgets covering latest reported information from fiscal years 2020 to 2022. In certain instances, current forestry budget was not included in the respective country annual budget. We obtained forestry expenditure information in these cases from the appendix to the UN FAO Global Forest Resources Assessment 2020 report, Economics and Livelihoods. The latest data available in this publication was from 2010. In these cases, we indexed the reported number to 2021 U.S. dollars using the individual country GDP growth rates. (Table created by Crowbar Research Insights LLC™ with information extracted from multiple sources. These sources are included at the end of the References section, © Crowbar Research Insights LLC™. All rights reserved.)

We tabulated this chart using published annual budgets for each country for the most recent fiscal year available between 2020 and 2022. Sources for individual country department budgets vary. Wherever possible, we used the forestry department budget contained within the overall country budget. In many instances, individual forestry department budget information is not published. Instead, we used the last UN FAO Economics and Livelihood report and indexed the reported forestry spending to 2021 dollars (U.S.) based on individual country GDP growth rates. While our methodology may not assure complete comparability, the information in this chart can serve as a guide to the financial resources each country commits to managing its forests.

Interestingly, Australia, China, Finland, Russia, and Spain commit the most resources to forest management. Germany has one of the lowest incidences of wildfires yet spends more on forest management than either Indonesia or Portugal, both of which incur significant yearly wildfire destruction and economic damage.

More noteworthy, the United States spends only 0.24 percent of its yearly federal budgeted expenditures on forest management and wildfire suppression. Australia, Canada, China, Russia, and Spain spend more. Further, the United States spends only marginally more than countries with less fire-prone geographies, such as Germany. This suggests the United States' investment in its forests certainly falls below what other countries with significant wildfire incidence spend.

FORESTLAND MANAGEMENT DIFFICULTIES

Deforestation describes the combined impact of wildfires, agriculture expansion, cattle ranching, mining, and logging on forest destruction. Such constant destruction threatens biodiversity, decreases carbon absorption, magnifies natural disaster damage, and disrupts water cycles.

According to the UN, in the decade ending in 2019, deforestation resulted in approximately 12 million acres of global forest loss each year.[70] Forestland management practices vary from region to region. Some countries ignore the significant ecological benefits prescribed burns can bring to forests while others employ little forestland management practice. Other countries decimate their forests through timber harvesting and logging.[71] The figure below summarizes annual forest change by continent since 1990 (see fig. 7).

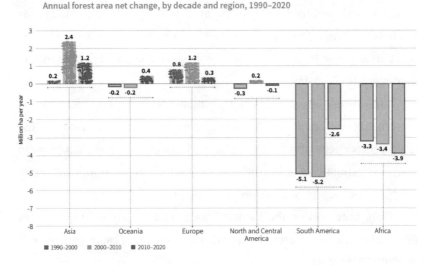

Figure 7. Annual forest area net change, by decade and region, 1990–2020. (Figure from the Food and Agriculture Organization of the United Nations, 2020, https://www.fao.org/forest-resources-assessment/en/.) [Figure credit: Food and Agriculture Organization of the United Nations, 2020, FAO, Global Forest Resources Assessment 2020, https://www.fao.org/forest-resources-assessment/en/. Reproduced with permission.]

International organizations, such as the New York Declaration on Forests (NYDF), attempt to address deforestation head-on through membership of more than 200 governments, multinational companies, groups representing Indigenous peoples, and nongovernmental organizations. These members commit to achieving ambitious targets for ending natural forest loss and restoring them. In addition, the UN,

WWF, and Boston Consulting Group are among those regularly publishing call-to-action reports.

In its November 2020 progress report, NYDF describes severe concerns about the growing extractive industries and planned infrastructure developments, particularly in developing countries. It expresses alarm over China's Belt and Road Initiative, which promotes massive infrastructure and mining activities in an estimated 126 countries, including already deforested Indonesia, South America, and Africa. The projects extend across sensitive ecosystems and biodiverse landscapes, establishing economic corridors and permanently removing intact forests.[72]

Forests with high ecosystem integrity provide greater forest benefits, such as carbon sequestration and storage, healthy watersheds, traditional forest use, contribution to local and regional climate processes, and forest-dependent biodiversity. Deforestation causes degradation to these forest benefits. A study published in 2021 reports that *only* 40 percent of the world's forests have high ecosystem integrity.[73]

Scientists estimate that globally, just 25 percent of wildfires are started naturally by lightning. Humans are responsible for the remaining 75 percent.[74] Deforestation and poor forest management practices are chief contributors to the increases in wildfire and international devastation.[75] Globally, higher temperatures increase the number of days wildfires are likely to burn.[76] In some cases, these fires occur in areas not commonly known for wildfires, leading to detrimental effects on wildlife and vegetation.[77] According to the WWF, "Southeast Asia does not have typical fire landscapes with naturally occurring wildfires, and thus the vegetation is not adapted to fire."[78]

A recent recommendation paper on wildfire management in Europe suggested insufficient fire suppression to combat wildfires. The CMINE Wildfire Task Group advocated for "a more pro-active fire management, where fuel loads are actively managed and fire strategies planned across landscapes. The benefits of such pro-active wildfire management are numerous: more controlled fires improved human health and decrease of economic losses." The task group cited higher temperatures, increased levels of forest fuels, and poor land management as significant contributors to European wildfires.[79]

Germany's forest management policy provides a guideline worth emulating: "Only through responsible and sustainable forest management will we be able to strengthen forests and ensure that they can continue to provide their many benefits for nature and people . . . Over time, people realized that a forest is not only a source of timber. A forest also protects soil, stores water, offers employment opportunities, and serves as an economic driver. A forest provides a habitat for plants and animals, a place for recreation, an air filter, and a source of oxygen. It contributes significantly to climate change mitigation."[80]

THE KEY TAKEAWAYS FROM THIS CHAPTER ARE:

- Every continent sustains significant human and economic losses each year due to wildfires, but the United States suffers more death and economic damage than any other country or continent.

- Wildfire and forest management practices develop slowly in every continent, including Europe and North America.

- Wildfire activity is rising across the globe, even in locations where they rarely occur.

- Deforestation negatively and significantly impacts forest ecosystems on every continent, particularly in Africa and South America.

INTERNATIONAL WILDFIRE AND FORESTLAND MANAGEMENT OBSERVATIONS

- Aerial firefighting represents a significant and impactful wildfire suppression strategy in Australia, Spain, Portugal, France, Italy, and Greece; however, aerial fighting in Africa, South America, and Indonesia has not developed due to several factors, including the lack of a designated government agency to coordinate regional, national, and continental resources.

- Politics and public awareness play an essential role in bringing together resources to better manage threats to forest ecosystems and from those perpetrating the damage.

- Fire suppression alone is an ineffective forest management practice. Forests managed through prescribed burns, forest fuels clearing, and replanting with mixed tree species appear hardier and more resilient to wildfires.

- The United States' annual investment in its forests lags far behind comparable countries with similar annual wildfire incidents. The United States also has the most to gain by increased investment in its forests to help reduce deaths and economic damage.

CHAPTER 9

REVIEW OF FORESTLAND MANAGEMENT PRACTICES

The cultivation . . . and maintenance of forests requires skill, knowledge, and commitment so they can be used permanently, stably, and sustainably.

—*Hans Carl von Carlowitz, 1713*[1]

According to the United Nations (UN), in 2020, forests covered 31 percent of the global land area or 10.3 billion acres. Forests are not equally distributed around the globe; two-thirds are found in just 10 countries with Brazil, Canada, China, the Russian Federation, and the United States containing more than half of the world's forests. Canadian and American forests make up more than 1.6 billion acres or 17 percent of the global forestland (see fig. 1).[2] The Canadian boreal forest is the largest intact forest ecosystem in the world.[3]

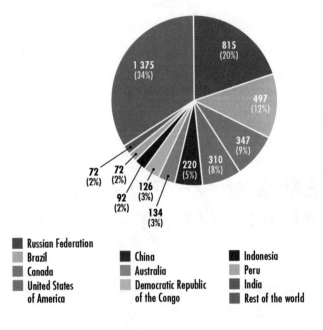

Figure 1. Global forest distribution shows the 10 countries with the largest forest area, 2020. (Figure by the Food and Agriculture Organization of the United Nations, 2020, http://www.fao.org/state-of-forests/en/.)

Forests are classified as either *primary* or *planted*. As defined by the UN, *primary forests* represent "naturally regenerated forests of native tree species, where there are no clearly visible indications of human activities, and their ecological processes are not significantly disturbed. They are sometimes referred to as old-growth forests." Approximately 93 percent of the world's forests are primary forests. These forests have extraordinary value for their biodiversity, carbon storage, wildlife species, regenerative capabilities, and other ecosystem benefits, including cultural and heritage significance. Sixty-one percent of such forests are found only in tropical and boreal regions—Brazil, Canada, and the Russian Federation.[4]

On the other hand, according to the UN, *planted forests* "can resemble natural forest at stand maturity and include forests established for ecosystem restoration and protection of soil and water." Seven percent (726 million acres) of all forests around the globe are planted forests; about 45 percent of these are plantation forests, comprised of one or two native or exotic tree species of similar age, planted with regular spacing, and intensively managed for productive uses.[5]

LOCATIONS AND CHARACTERISTICS OF AMERICA'S FORESTS AND LAND COVER

Approximately 33 percent of the U.S. landmass is forestland (see fig. 2). Of the 746 million acres of forests in the United States, 92 percent (686 million acres) are primary and naturally regenerative forests, the most biodiverse and carbon-dense type of forest. There are 60 million acres of planted forests, with about 3 million acres representing plantation forests located mainly in southern states.[6]

Figure 2. Land cover classifications in the United States. (Multi-Resolution Land Characteristics Consortium, 2019, https://www.mrlc.gov/data/legends/national-land-cover-database-2019-nlcd2019-legend.)

About two-thirds (514 million acres) of the U.S. forests are considered timberland. These forests can produce 20 cubic feet per acre of wood used for industrial purposes annually.[7] Another 52 million acres of forest, such as parks or wilderness areas, are reserved for non-timber uses and are managed by public agencies. Additionally, 191 million acres of other forestlands are not capable of producing 20 cubic feet per acre of industrial wood annually, but of significant importance for watershed protection, wildlife habitat, domestic livestock grazing, recreation, biodiversity maintenance, and gathering nontimber products such as berries, mushrooms, and medicinal plants.[8]

Figure 3. Forest types across the 48 contiguous U.S. states and Alaska, derived from both a forest type group map of the contiguous United States and a forest type group map of Alaska. (Springer Open, 2015, https://forestecosyst.springeropen.com/articles/10.1186/s40663-015-0045-4.)

Four federal agencies are responsible for and are stewards of public forestland (USDA Forest Service, Bureau of Land Management, National Park Service, and the U.S. Fish and Wildlife Service), along with numerous states, tribal, county, and municipal government organizations.[9] About 10 million private individuals and companies own approximately 422 million acres of forest and other wooded lands.

The image below displays the forest type by geographic region (see fig. 3). Each area has different types necessitating unique management and fire suppression practices. Techniques that might work on Maine aspen and birch forests will unlikely be effective for Pacific Northwest Douglas fir. Each region is unique and so are its forest productivity, density, wildlife inhabitants, biomass, carbon storage, and species diversity. Wildfire suppression tactics, likewise, require tailored execution based on regional conditions.

Understanding landmass characteristics, biodiversity-ecosystem interactions, and ownership represents the fundamental pillars of any forest management undertaking. Fortunately, our research indicates no shortage of information about these pillars. Arguably, the plethora of research, papers, conferences, journals, studies, trade association materials, congressional commissions, and government publications confuse and obfuscate the execution of consistent, sound forest management practice.

DEFORESTATION AND DEGRADATION

Deforestation. In 1630 before the arrival of Europeans to the Americas, about one-half of the U.S. land area was forest or about 1.02 billion acres.[10] Most deforestation in the United States took place before 1910 with the settlement of the west. In 1920 the USDA Forest Service (FS) reported forestation as 721 million acres around 1920, and since then, forests have remained constant.[11, 12] Nonetheless, a 2017 study estimated a 3 percent loss of forests between 1992 and 2001 due to wildfires and insect infestations.[13] Between 2010 and 2020, forests increased 0.30 percent increase in the U.S. forests, partially reversing the trend of previously reported losses. This increase resulted from forestland regeneration.

Principal causes for deforestation in the United States are wildfires and insect infestation resulting in significant forest fragmentation and Wildland-Urban-Interface (WUI), defined as urban expansion that obstructs sensible forest management tactics. Forest fragmentation induced by wildfire, pests, and WUI cause segmentation of the forest into smaller patches or combustible trees and underbrush, thereby creating the fuels for what will likely be intense wildfires. The photograph below, taken in 2021 along the I-70 corridor in Colorado from Denver to Vail is one example of fragmentation caused by uncontrolled mountain pine beetle infestation (see fig. 4). A lightning strike of kindling of damaged and dead trees will ignite these trees and possibly spread the conflagration to adjoining forests. It is a real threat that has increased in the last thirty years, yet there is little, if any, insect mitigation effort by the FS or the Bureau of Land Management (BLM), the owners, and the managers of this forestland.

Figure 4. There are extensive mountain pine beetle-damaged forests (brownish-gray trees) along the I-70 corridor between Denver and Vail, Colorado. (Photo by author, July 2021.)

Bark beetles, mountain pine beetles in the west, gypsy moths, Asian long-horned beetles, emerald ash borers, and woolly adelgids in the east and mid-west have brought about extensive damage and death to trees in U.S. forests, infestations responsible for almost $900 million in estimated costs each year.[14] Moreover, the UN estimates insect the extent of pest damage at 86.5 million acres of global forests each year.[15]

Bark beetle and other invasive species attacks on trees result in an increase in flammability due to a decrease in moisture. Eventually, the tree branches fall, the trees topple, and the undergrowth biomass drastically increases creating fuel for more intense wildfires. Research in the Rocky Mountains determined that conifers attacked by bark beetles and then exposed to a wildfire cannot recover from this double destruction.[16]

There is controversy over bark beetle infestations and their role in catastrophic wildfires. Some scientists believe climate change is the root cause of the uptick in beetle outbreaks and more destructive wildfires. Others argue the link between beetle infestations and intensifying wildfires is not conclusive.[17, 18] Since 2001, legislators have introduced more than 50 bills to increase timber harvest supposedly to reduce bark beetle infestations. More recent research suggests that some wildfire management practices (thinning and prescribed burns) may improve tree health, increase beetle resistance, and promote tree growth.[19]

> ***Authors' Note***
> Beetle damage control efforts began in 1906 in the Black Hills National Forest in South Dakota. In 1930 a proposal at an anticipated cost of $1 million was introduced to eradicate bark beetle infestation.[20] The plan remained just that—a plan. Instead of eradicating, the FS chose to continue to study the problem. Years later, a 2008 article concluded bark beetle outbreaks and the interaction between beetle infestations and destructive wildfires still required additional research.[21] After 117 years, bark beetles continue to damage our forests with apparently no concerted effort to eradicate them.

Despite decades of study and research, the FS and BLM have differing infestation policies. FS-funded studies offer no conclusive infestation treatments, possibly because the controversy is either hung up in the federal bureaucracy or weighed down by—dare we say it—incompetence. Even worse, other studies question whether bark beetle destruction causes catastrophic increases in wildfires despite data and common sense to the contrary.

Degradation. The UN defines *forest degradation* "as a reduction in the biological or economic productivity of forest ecosystems resulting in the long-term reduction of benefits from the forest, which includes wood, biodiversity, carbon sequestration, and recreation."[22] The lines between deforestation and degradation can quickly become blurred, particularly in the United States. Unfettered undergrowth in our national forests for the past 117 years is an unchecked opportunity for intense wildfires. The longer government agencies and congressional committees defer appropriations to clear the woods, the longer and more intense fire seasons will be. Some argue that wildfire suppression is counterproductive due to fallen trees and debris from their decay. They use this rationale for "let burn" practices to clear the debris from forest floors as the only forest management practice available.

FS AND BLM FOREST MANAGEMENT PRACTICES

Federal wildfire management practices have changed over the more than 117 years since the 1905 creation of the FS. One of the first policies, initiated in 1926, was the "ten-acre policy" that stated, "all wildfires should be controlled before they reach ten acres in size." Next came the "10:00 a.m. policy" in 1935, saying, "any fire exceeding ten acres should be controlled before 10:00 a.m. the day after the initial fire report." In the 1970s, the FS and the National Park Service (NPS) implemented "prescribed natural fire" policies, where fires were monitored and allowed to burn rather than suppressed unless the fire exceeded the prescribed area. "Let burn" policies ceased after the disastrous 1988 Yellowstone National Park wildfires, and suppression again became the primary method of combating wildfires. In 1995 the FS revised the practice due to the deaths of several firefighters in severe wildfires the previous year.

Current policy is "to protect human life first, and then to protect property and natural resources from wildfires," apparently to be accomplished through wildfire suppression and continued use of "let burn" policies.[23,24] According to the National Association of State Foresters, "Natural forest disturbances—whether caused by pests, severe wind, lightning, or other means—change the structure and composition of forests and allow for regeneration. Many forest types need one natural disturbance in particular to regenerate, and that's fire."[25]

The role of Congress is to appropriate funds for federal wildfire and forest management, suppression and fuel reduction (eliminating vegetation and thick undergrowth from forests), preparedness (wildfire forecasting, equipment and resource management, workforce training), and other wildfire activities such as wildfire research. The FY2020 wildfire suppression appropriation was $6.11 billion, which is the highest budget to date; however, the funding missed its goals.[26]

Maintenance and restoration fall under the forest management umbrella. Management costs include public safety, education, and research to develop better management and forest improvement techniques. Clearing underbrush, prescribed burns, and tree thinning to sustain the forests are examples of activities to provide maintenance for our forests. These are referred to as forest management costs throughout the literature we researched.

In 2017 Agriculture Secretary Sonny Purdue stated, "Forest Service spending on fire suppression in recent years has gone from 15 percent of the budget to 55 percent—or maybe even more—which means we have to keep borrowing funds that are intended for forest management." Purdue attributed this increase to longer than average fire seasons.[27] We can reasonably conclude programs like wildfire prevention and preparedness along with other forest management programs are severely underfunded. The accuracy of Secretary Purdue's comments is illustrated below (see fig. 5).

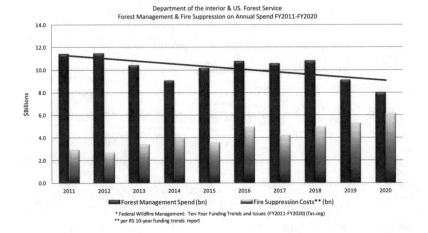

Figure 5. Federal agency forest management and wildfire suppression costs. (Figure created by © Crowbar Research Insights LLC™. All rights reserved.)

As figure 5 shows, fire suppression costs at the federal level represented about 45 percent of the total 2020 federal forestry budget, a 25 percent increase above 2011. The trend line indicates a probably high funding risk for future forest management activities adding to the already dangerous deferred forest maintenance situation.

Suppression is the primary practice for combating wildfires through such efforts as fuel reduction, timber management, and extinguishing wildfires (see chapter 3). Mechanical thinning equipment includes bulldozers, mowers, chainsaws, masticators, and other equipment to cut and remove vegetation to reduce the risk of severe wildfires. Removal of forestland underbrush—that is, fuel—consists of mechanical thinning and prescribed burns to reduce tree density and thick undergrowth. As discussed in chapter 5, we found little evidence of year-round training on mechanized equipment to assist in reducing dead trees and forest underbrush. Prescribed burns, also known as controlled or prescribed fires, are deliberately set fires to reduce fuel loads (see fig. 6).

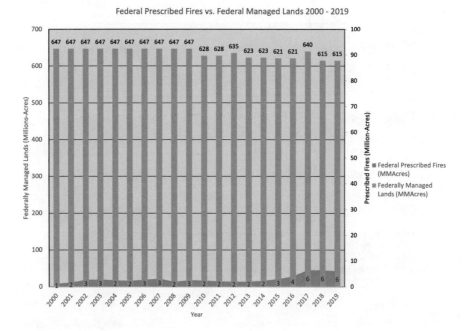

Figure 6. Annual prescribed burns versus federal lands managed—2000 to 2019. (Figure created by Crowbar Research Insights LLC™ with data from National Interagency Fire Center Fire information, https://www.nifc.gov/fire-information/prescribed-fire, © Crowbar Research Insights LLC™. All rights reserved.)

Burn plans are created (prescribed) based on outdoor conditions (wind, humidity, temperature) and the safety of the public and fire staff. The practice of prescribed burns increased during the last few decades, with a 28 percent increase in the number of acres treated with prescribed burns in 2019 compared with 2011. In the United States, more than 10 million acres of forestry and rangeland were treated with prescribed burns in 2019.[28] But our charts reveal these prescribed burns barely put a dent in treating total acreage under federal agency management (see figs. 7 and 8).

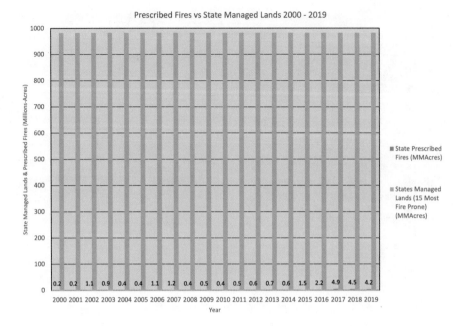

Figure 7. Annual prescribed burns versus federal lands managed—2000 to 2019, in the fifteen most fire-prone western states. Acres managed include federal and state lands in the 15 most fire-prone states described in chapter 4. (Figure created by Crowbar Research Insights LLC™ with data from National Interagency Fire Center Fire information, https://www.nifc.gov/fire-information/prescribed-fire, © Crowbar Research Insights LLC™. All rights reserved.)

State and local air quality managers frequently invoke their regulations to limit the number of prescribed burns based on local conditions. As a result, prescribed burns play a minimal role in today's U.S. forest management practices.

Timber harvesting, often called logging, is the process of cutting down, removing, and selling trees to support the health of the forest and reduce wildfire risk. Each year, the FS and the BLM conduct timber sales with hardwood harvested on federally managed land. Harvesting is permitted but rarely performed on land managed by the NPS and the U.S. Fish and Wildlife Service (FWS). The FS reports it can treat 17 million acres with timber sales and 35 million acres by mechanical thinning and prescribed burns (see figs. 8 and 9).[29, 30]

REVIEW OF FORESTLAND MANAGEMENT PRACTICES

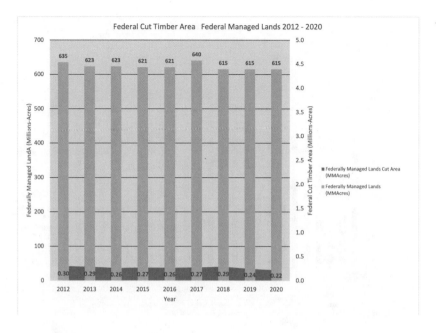

Figure 8. Federal acres timber cut 2012 to 2019. (Figure created by Crowbar Research Insights LLC™ with data from the Forest Service website: Reports and Publications, https://www.fs.fed.us/forestmanagement/reports/index.shtml and https://www.fs.fed.us/forestmanagement/documents/harvest-trends/NFS-harvesthistory1984-2020.pdf, © Crowbar Research Insights LLC™. All rights reserved.)

The Oregon and California Railroad and Coos Bay Wagon Road Grant Lands Act of 1937 (O&C Act) established 2.4 million acres of specific federal lands in western Oregon and California classified as timberlands were to be managed for permanent forest production, and the timber was to be cut, removed, and sold in conformity with the principle of sustained timber production yield. In 2020 a federal court ruled the BLM was not upholding the O&C Act, allowing for sustainable timber production.[31] This is an example of federal bureaucrats overriding laws enacted by Congress to the detriment of healthier forests and local residents' livelihoods.

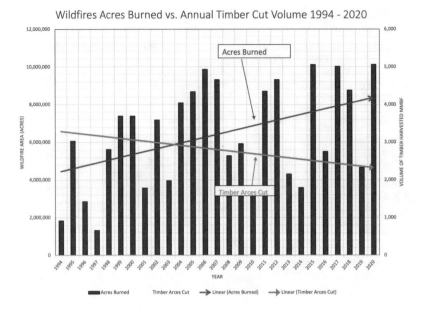

Figure 9. United States acres destroyed by wildfires versus federal acres timber cut 1994 to 2020. (Figure created by Crowbar Research Insights LLC™ with data from the Forest Service website: Reports and Publications-Harvest Trends on National Forest System Lands Historic Harvest Records, 1984 to present, https://www.fs.fed.us/forestmanagement/documents/harvest-trends/NFS-HarvestHistory1984-2017.pdf, and BLM, https://www.blm.gov/about/data/public-land-statistics, © Crowbar Research Insights LLC™. All rights reserved.)

Figures 8 and 9 verify the existence of three troubling concerns: (1) timber harvesting represents an insignificant impact on the nation's inventory (less than 0.04 percent); (2) timber board feet harvested in the past thirty-six years steadily declined; and (3) *there appears to be a correlation between reduced timber harvesting and increased acres destroyed by wildfires*. During this same period, federal agencies' receipts from harvesting lumber declined to just $65 million in 2020, a 75 percent drop from mid-1990 levels.[32] Why?

One reason is funding challenges. Since 2011, forest management spending has steadily declined to $8 billion in 2021, a 30 percent cut. We do not support timber harvesting for purely financial gain; however, *sustainable and responsible timber harvesting* represents an opportunity to reduce wildfire risk and provide funding to help replenish defrayed forest management budgets. Conventional timber sales, like other stewardship projects, can help the agencies achieve conservation and fuels reduction goals and generate revenue from public lands, while supporting rural jobs and funds for county services.

SHARED STEWARDSHIP AGREEMENTS

The Good Neighbor Authority (GNA), passed by Congress in 2014 and expanded in 2018, authorizes government agencies to partner with tribes, state, and county agencies to implement coordinated cross-boundary forest management projects. For its 2021 FY justification, the FS reported 179 active GNA agreements: among them 67 agreements for forest health, 28 for wildlife management, 27 for timber management, and 18 for fire management. Most notably, none of these agreements was with a federally recognized tribe. In 2019 the BLM, which reports its GNA data differently than the FS, entered 26 new GNA agreements. It is important to note that GNAs are not funded separately within the FS or BLM budgets. Instead, these agencies can use any available funds for them.[33]

Building on this presumed cooperation and collaboration, the FS created a Shared Stewardship Program authorized through the Consolidated Appropriations Act of 2018 (the omnibus bill) that expanded GNA and extended stewardships to twenty years. This stewardship program is one of five FS priorities. "A third priority for the Forest Service is promoting shared stewardship by increasing partnerships and volunteerism," stated former FS chief Tony Tooke in 2017.[34] This program focuses on all the wildlands in the United States in addition to FS-managed areas. Its goals are:

- improve the condition of lands, forests, and grasslands through collaborated investments with states and tribal nations;

- work on projects having the most impact to improve forest health and resiliency across management jurisdictions; and

- use science and previously proven outcomes as critical components for the activities undertaken.

The FS currently has 33 shared stewardship agreements with states or institutions. The FS and Tribal Nations (Tribes) created the Tribal Relations Strategic Plan to increase engagement in shared stewardship of lands where Tribes retain legal rights and interests along with 4,000 miles of shared boundaries.[35-37]

Since 2003, the BLM has issued at least 170 stewardship awards under Public Law 108-7 granting stewardship contracting authority to tribal nations and private companies. These agreements create a value for timber and vegetative products removed to offset the cost of the services for removal.[38]

Shared stewardship agreements make good headlines. They contain sound forest management ideas and facilitate better coordination for federal, tribal, and state forest management agencies. What's missing? None of the shared stewardship agreements provide any meaningful funding to accomplish the FS and BLM goals outlined in these agreements. As with so many FS and BLM initiatives, our work concludes that the bureaucratic apparatus in trying to appease public concern announces its latest programs instead of actively advocating for funding to implement proven solutions for forest management issues.

NATIVE AMERICAN FOREST MANAGEMENT PRACTICES

Native Americans are the original stewards of U.S. forests and wildland. For many Native Americans, forests were—and still are—a crucial resource, providing food, shelter, culture, medicine, and employment.

Tribal fire specialists, also known as "burners," conduct prescribed burns for crop management, insect control, land clearing, and promoting healthy growth. They do not suppress natural wildfires. Native Americans manage their lands for future generations and engage in maintaining and growing healthy forests.[39]

In the last 150 years, government agencies have taken over this responsibility. The federal and state bureaucracy controls about 19 million acres of Indian forests and woodlands. Government control is slowly changing as federal agencies accommodate Native Americans' emphasis on protecting their valued forests. Legislation, including The National Indian Forest Resources Management Act of 1990, the Tribal Forest Protection Act of 2004, and the Agriculture Improvement Act of 2018 (known as the 2018 Farm Bill), recognize the beneficial methods used by Native Americans to manage their forests.[40]

Yosemite National Park in California is a prime example of those changes. Historically, Native Americans burned Yosemite Valley periodically to keep the meadows open, encourage regrowth and maintain forests. In the late 1800s, early settlers initiated fire suppression and exclusion practices continued by Yosemite Park rangers. Open views within Yosemite were once hindered by encroaching vegetation. Instead, the NPS uses prescribed fires and thinning to manage forests allowing nature-caused wildfires to burn without suppression except when they threaten people or property.[41, 42]

The following photographs compare vegetation in Yosemite Valley under indigenous people management in 1866 and FS management in 2013 (see figs. 10 and 11). Notice the stark increase in tree growth and density due to government agencies' lack of active forest management. Before European Americans' settlement, Native Americans in Yosemite had a long tradition of periodically burning parts of Yosemite Valley and other meadows with other traditional land management practices. Periodic burns kept the meadows open, encouraged indigenous foods and materials regrowth, and maintained the California black oak woodlands. Fire suppression and fire exclusion were begun by early settlers in the late 1800s and continued by FS managers for decades. These practices replaced many original black oak woodland areas with aggressive, shade-tolerant coniferous species, such as incense cedar and white fir. Forest undergrowth increases when these species take hold.

Figure 10. Yosemite Valley, View from Union Point, 1866. (Photo by Carleton E. Watkins, 1866, https://oac.cdlib.org/ark:/13030/kt6d5nc840/?brand=oac4.) [Photo credit: Courtesy of Yosemite National Park, Carleton E. Watkins, 1866, YOSE 9527.]

Under the Tribal Forest Protection Act, tribes could propose specific forest management projects. The Good Neighbor Agreement allows tribes to conduct management projects via cooperative agreements with federal partners. To date, only three tribes have engaged in GNA agreements.[43] The 2018 Farm Bill included 60 tribal priorities, including the

Tribal Forest Management Demonstration Project, which authorizes the U.S. Department of Agriculture and the Department of the Interior to enter into contracts with tribes and tribal organizations to permit tribes to manage and protect forests from fires and otherthreats.[44]

Figure 11. Yosemite Valley, View from Union Point, 2013. (Photo by the National Park Service, 2013, https://npgallery.nps.gov/YOSE/GetAsset/ d2757b52ef29455690d6b61f96cc4987/proxyhires.)

The federal government selectively adopted several forest management techniques used by the Menominee Tribe of Wisconsin. The Menominee made a living selling lumber and created a tribal mill on their land. The tribe engaged in selective cutting of older trees, preferably those older than 200 years, instead of clear-cutting their forests and relying on rotation ages of 80 to 100 years. The tribe initiated a tree-planting program consisting of various trees to replenish the land instead of planting only fast-growing trees. The Menominee suggested the Continuous Forest Inventory program, which created permanent plots to allow the Bureau of Indian Affairs (BIA) to estimate average growth on the reservation. This long-term monitoring helped determine a sustainable cut level in the forests. Conducted in 1993, 2003, and 2013, a national study required by the National Indian Forest Resources Management Act (NIFRMA) concluded "Indian people tended to prefer 'protection' and the use of forests for 'subsistence.' The non-Indians in the BIA believed their clients preferred income and the use of the forests to provide income." The two groups came together by the 2003 report and emphasized the importance of Indian values.[45]

The Coquille Indian Tribe of Oregon uses a longer harvest cycle to manage Douglas firs prominent in the northwest, including the Coquille forests red and Port Orford cedar trees. The Port Orford Cedars only grow in a 30-mile-wide band of the coastal regions in southern Oregon and northern California. The tribe inventories the forest to ensure a sustainable timber harvest and thin 20–40-year-old trees, but, when necessary, it clear-cuts areas of their forest while they protect nests and other wildlife habitats.

"The Coquilles are the only forest managers in their neighborhood who are meeting both their timber-production and their environmental targets. They are committed to long-term, environmentally sound forest management," according to John Gordon, an emeritus forestry professor from Yale. The forests of both the Coquille and Menominee Tribes are certified under the Forest Stewardship Council, a worldwide organization ensuring forests are economically, socially, and environmentally managed.[46]

The 2013 report required by NIFRMA found that forest health was better on tribal land than on the surrounding federal forestland. Our research found many other examples demonstrating superior forest management stewardship by Indigenous peoples to those practices imposed by federal agency bureaucrats. That should be more than enough notice to the federal bureaucracy and state and local agencies that it is time to listen and learn from the original stewards of our forests.[47]

OTHER SUCCESSFUL FOREST MANAGEMENT RESTORATION PROJECTS, PRACTICES, AND ORGANIZATIONS

Forest management practices establish goals of managing and protecting the health of a forest from threats including wildfires, insects, diseases, human interference, and climate change. A significant component is control of vegetation within the forest and restoration of ecosystems after a wildfire, drought, or flood. Practices focus on reforestation including providing and planting seeds from genetic resource programs, nurseries, and seed extractors. Silviculture treatments (thinning, pruning, growing) decrease tree mortality and promote growth, regeneration and restoration along with timber harvests and prescribed burns for wildfire management. Inventorying the forests to assess their health and status provides critical data for planning forest and wildfire management practices.

The French Meadows Restoration Project is a successful example of forest restoration and an important private-public partnership project. In 2014 the King Fire burned more than 97,000 acres in the Eldorado National Forest and on private lands in California. The wildfire damaged the Placer County Water Agency's (PCWA) French Meadows and Hell Hole Reservoirs followed by postfire erosion that continued to affect the reservoirs and PCWA's downstream infrastructure. PCWA joined the FS and several other private organizations to form the French Meadows Partnership with a goal of restoring forest health on more than 12,000 acres of federal and private land (see fig. 12). The area was already on the FS planning schedule, but resources were unavailable to move forward with any forest management projects prior to the partnership.

Figure 12. French Meadows Reservoir is in the upper reaches of the American River watershed. The Sierra Nevada provides more than 60 percent of California's developed water supply, and clean water is one of the most critical benefits provided by forested watersheds. (Photograph by © Placer County Water Agency, May 2019. Used with permission.)

In 2016 the partners executed a Memorandum of Understanding to build, support, and raise funds to implement restoration tasks. Through December 2020, they treated more than 3,100 acres primarily through hand-thinning methods due to the remoteness of the forest. Their next project is aspen and meadow restoration in an area exceeding 12,000 acres. A Master Stewardship Agreement between Placer County and the Tahoe National Forest authorizes mechanical thinning, mastication, hand thinning, and aspen and meadow reforestation for restoration from damages. In addition, the project involves more than 7,000 acres of prescribed burning, to be managed by the Nature Conservancy and the FS.[48]

In 2020 the partnership brought more than 1.4 million board-feet to a local mill and at least 1,200 tons of biomass to a local renewable energy facility. The sale of wood products and biomass helps fund the project.

With a $12 to $14 million budget, this project's expected cost is approximately $110 per acre. Another key finding of this partnership is significantly reducing "the typical time for planning forest restoration projects on FS lands while also reducing the burden on limited federal staffing and resources.[49-51]

Another cooperative project is Tree Cities of the World, an international effort promoted by the Arbor Day Foundation and the UN. Its goal is to encourage and recognize cities and towns committed to maintaining and sustainably managing their urban forests. Written statements from city managers delegate responsibility for tree care to a tree board within a municipal boundary. The board adopts policies, best practices, and industry standards to manage urban trees and forests. The tree board has an annual budget to implement best practices and imposes penalties and fines for noncompliance.[52]

The National Association of State Foresters represents state foresters who manage and protect state and private forests, roughly two-thirds of the nation's forests. The association advocates for federal legislation and national policies to promote the health, resilience, and productivity of forests across the country on behalf of the professionals who conserve, enhance, and protect our forest resources. The organization issues thoughtful, balanced policy statements on forest management, biomass and renewable energy, endangered species reform, invasive species management, and other timely subjects focused on achieving a balanced set of economic, environmental, and social benefits.[53]

State forestry departments also develop creative solutions to wildfire risks. The Texas A&M Forest Service, working with its state legislature and casualty insurance industry, created a public-private partnership to fund the acquisition of fire equipment, protective gear, and training tuitions for the state's rural volunteer fire departments. Since 2002, more than 40,000 requests totaling almost $330 million were funded allowing Texas rural communities to acquire modern equipment and provide training to volunteer forces. Casualty insurance companies provide the funds to the state and as a result, the insurance companies experience a lower volume of claims due to investments in equipment and training of volunteer forces for fighting fires more effectively throughout Texas's vast private land areas. South Carolina adopted a similar solution.

New York Declaration on Forests (NYDF) is a political declaration calling for global action. It offers a standard, multistakeholder framework for forest action, consolidating various initiatives and objectives to drive forest protection, restoration, and sustainable use. Adopted in 2014, it received endorsements from over 190 entities to date, including more than 50 governments, more than 60 of the world's biggest companies, and more than 80 influential civil society and indigenous organizations. The NYDF is a significant reference point for global forest action. Its ten goals include halting natural forest loss by 2030, restoring about 864.9 million acres of degraded landscapes and forestlands, improving governance, increasing forest finance, and reducing emissions from deforestation and forest degradation.[54]

United Nations State of the World Forests 2020 sets forth in its report a series of recommendations to reverse global forest loss and degradation through sustainable forest management practices. It encourages balanced solutions, using localized, regional resources to develop policies and actions under a governance structure accountable to local populations. It also promotes innovative financing to accomplish local forest management goals.[55]

ISSUES WITH WILDFIRE MANAGEMENT PRACTICES

Wildfires are not a new phenomenon. By analyzing fossil charcoal, researchers estimated wildfires first occurred 420 million years ago after the appearance of terrestrial plants.[56] There are numerous benefits to controlled fires, such as warmth, light, heating, cooking food and water, and keeping insects and predatory animals away. Native Americans and early settlers used fires to clear areas to stimulate grass growth and create agricultural regions.[57]

Surprisingly, there also are benefits to wildfires. They clear out underbrush allowing more sunlight to penetrate the forest and nourish the soil resulting in new and regenerated plant life for food and shelter for wildlife. Wildfires kill diseases and insects that destroy trees. Several tree and plant species require the heat from a fire to release seeds for regeneration or have adapted to living in areas burned by wildfires.

In December 2019, the U.S. Government Accountability Office (GAO) issued a report titled *WILDLAND FIRE—Federal Agencies' Efforts to Reduce Wildland Fuels and Lower Risk to Communities and Ecosystems*. The report states wildfire management practices of federal agencies "altered the normal frequency of fires in many forest and grassland ecosystems."[58] The following paragraphs summarize critical parts of the GAO report related to the challenges and concerns with federal wildfire management practices.

SUPPRESSION

As mentioned earlier, suppression practices include fuel reduction (mechanical thinning and prescribed burns), timber management, and extinguishing or suppressing wildfires. The primary means is to extinguish them quickly, but this practice can increase dense undergrowth, sometimes called biomass, comprised of dead or dying trees and small thick trees, shrubs, and other plant life areas. In addition to contributing to wildfire intensity—a point that cannot be overemphasized—undergrowth can facilitate insect and disease infestations that can kill more trees.[59, 60]

While mechanical thinning is effective, it too has downsides. First, it may increase slash or leftover debris such as bark, branches, and brush. Slash can increase the intensity and rate of wildfire speed. Second, previously thinned areas require ongoing maintenance as undergrowth returns. Timber harvests also can increase slash.[61, 62] But recent advances in technology reduce these residual impacts.

Prescribed burns are efficient at reducing smaller fuels but not larger ones due to the risk of the fire spreading outside of the defined area. In addition, smoke from the burn affects air quality contributing to air pollution.[63, 64] Interestingly, a comprehensive review of fifty-six studies on the effectiveness of fuel treatments concluded that thinning or prescribed burns alone had little to no effect on reducing wildfire severity but thinning used in combination with prescribed burns had a positive impact on lowering it.[65]

In response to the GAO's 2019 report, agency officials stated their staff was occupied with wildfire suppression and unavailable for other forest management projects.[66] Agencies claim they do not have enough staff to treat the areas they manage plus wildfire suppression costs consume their budgets, forcing them to divert funds from other wildlife and forest management activities ("fire borrowing") or seek emergency appropriations.

The 2018 omnibus bill provides federal agencies access to a disaster fund, known as the "fire funding fix," allowing them to pay for wildfire suppression costs without borrowing funds from their other activities. Through 2027, the FLAME account (see chapter 2) authorizes an initial funding level of $2.25 billion. The fund intends to free up about $1.3 billion annually for agencies to protect their lands—that is, thinning and prescribed burns—from catastrophic wildfires.[67]

Federal agencies clearly are struggling with funding their wildfire management practices. According to the Worldwide Fund for Nature and the Boston Consulting Group, "[T]he fire management budget has been cut from $6 billion to $4.75 billion, while the focus on suppression increased from 11% to 55%.[68] In 2015 the FS calculated that it spent 30 percent or more of its budget fighting only one to two percent of fires. By 2025, fire suppression costs will increase substantially.[69]

WUI DRIVEN BY HOUSING EXPANSION

The WUI is the area of land where human development encroaches or intermingles with undeveloped wildland vegetation. WUI places communities and infrastructure at increased wildfire risk. WUIs are mostly on non-federal land. While some state and local agencies require fire reduction mitigation activities—fire-resistant building materials and reducing vegetation on and around these properties—responsibility for following through with these critical strategies rests with individual property owners. According to a GAO report, 70,000 communities comprising more than 46 million single-family homes, plus additional structures, are in WUI and at risk. So are facilities and infrastructure such as power lines and roads not located immediately adjacent to WUI, due to wind-carried embers from wildfires. Between 2002 and 2016, an average of more than 3,000 structures in WUI areas were damaged or destroyed by wildfires yearly.[70, 71]

One severely underused strategy in wildfire management is regulating or limiting development in areas designated as WUI. As of the previous decade, approximately one-third of the U.S. population lived in a WUI area mostly in California, Texas, Florida, North Carolina, and Georgia. Surprisingly, the WUI area increases by 2 million acres per year. Unfortunately, due to increasing population growth, the rising cost of housing, the shortage of available and affordable housing, and a trend of moving into closer to nature, WUI encroachment ordinances are rarely enforced at the local, state, or federal level.[72]

While reducing vegetation in WUI can lower risk and intensity, federal agencies state fuel reduction projects in WUI areas are more costly. One estimate from a federal agency official projects costs of prescribed burns in WUI area communities at around $250 per acre compared to $60 per acre in non-WUI areas. The cost is a significant challenge in combating WUI wildfires because public perception appears to be that the risk to WUI communities is negligible.[73]

CASE STUDY—TAMARACK FIRE[74]

FOREST MANAGEMENT OR AGENCY NEGLIGENCE?

On July 4, 2021, an FS helicopter observed the outbreak of a small lightning-caused, quarter-acre fire in the Tamarack Forest in Mokelumne Wilderness in Alpine County, California, and Douglas and Lyon Counties, Nevada. The following link captures the event: https://fb.watch/9LU1WjVoUt/.

FS personnel made a "let burn" decision. Later explaining their decision, the FS justified its decision based on human safety issues and difficulty to access the fire due to the surrounding terrain.

On July 16, gusty winds expanded the fire, turning it into an uncontrolled burn. The Tamarack wildfire consumed 68,637 acres, destroyed 23 buildings, damaged three structures, and killed an unknown number of animals until finally extinguished on October 31, 2021.

Three advance air bases (Sacramento, Reno, and Bishop), two of which had single-engine air tankers and helicopter assets, were within a 60-to-100-mile striking distance. Nearby lakes could supply the water needed for Bambi-bucket/snorkel-equipped helicopters. FS personnel chose none of these options.

Rather than incur a cost of less than $100,000 to put out the fire, this wildfire ultimately cost an estimated $37 million, created smoke emissions for two months, denied public access to this wilderness area, and damaged the forest area.

CONCERNS WITH FOREST MANAGEMENT PRACTICES

As with the wildfire management practices, there are several issues with the current forest management practices related to lack of funding, agency negligence, insect and owl concerns, and ignoring assets.

INSUFFICIENT PHYSICAL ASSETS AND AGENCY MISMANAGEMENT

Federal agencies manage 615 million acres of land, representing 28 percent of the 2.2 billion acres of land in the United States.[75] Regrettably, hampered by availability of resources, these agencies' forest management practices embrace the status quo. They also carry out self-inflicting wounds, such as decommissioning 1,000 to 2,000 miles of forest roads every year thereby reducing access to wildfire incidents by firefighting teams. They do not receive the required funding to perform vital forest management practices; neither are they energized to change their self-serving behavior. Mechanized equipment, aerial firefighting assets, year-round forest maintenance personnel, supplies, and training opportunities go wanting.

Forest management agencies are no different from other federal agencies. Limited by bureaucratic red tape and regulations, congressional underfunding, and inevitable budget reductions, they must operate under continuing resolutions when appropriations bills do not pass by the start of the fiscal year. Some agencies delay hiring staff and implementing planned projects. The FS prefers to conduct prescribed burns in fall and winter, but delayed funding reduces the number they can actually conduct.[76] Lack of staff availability is a significant concern. In addition to the demand suppression efforts place on staff, since 1992, the number of skilled employees needed for treatment and restoration work fell by 54 percent, according to an estimate from the National Association of Forest Service Retirees.[77] Recently released statistics indicate the trend may continue. The FS budget for 2021 called for a fire season workforce of 10,000 personnel, yet due to a range of factors, including less than attractive pay scales, poor contracting processes, and a change in field offices responsible for hiring, only 75 percent of personnel budgeted were under contract at the commencement of the fire season in April 2021. FS personnel also state their goal to create a 10,000-personnel year-round workforce, yet only 1,000 such personnel were budgeted for fiscal 2022, leaving resources woefully inadequate to tackle the off-fire-season demand to clear, thin, and manage our forests.[78, 79]

Of the 100 million acres at risk from wildfires, the number of acres agencies actually can manage yearly is just three million acres—the proverbial drop in the bucket.[80] While the 1980 federal Reforestation Trust Fund provides some funding, only $30 million annually is available for national forest planting projects, limiting the agencies to about 15 percent of the 1.3 million acres in critical need of new planting.

According to a report to Congress in 2019 prepared by the Congressional Research Service (CRS), FS personnel and stakeholders identified administrative process barriers, funding, and compliance with government requirements as factors obstructing forest restoration. (The FS is the author of these processes.) Previous GAO reports found that about a fourth of FS fuel reduction decisions and a fifth of fuel management projects were challenged through appeals or objections. The CRS report concluded that agency management and administration have become "more complex and contentious," on issues of multiple land use and sustained yield. Where once the decision was the province of the FS, Congress increasingly is enacting legislation that strips decision-making from the agency and regulates land use and sustained yield.[81]

A fire technical specialist and a member of the Intertribal Timber Council, stated, "Unfortunately, national forest management is gridlocked by special interests, complicated federal laws, conflicting policies, and inconsistent court-imposed requirements. Instead of a consistent flow of well-planned projects implemented on a reliable schedule, our national forests are compelled to invest a sizable proportion of their resources to planning, analysis, appeals, and lawsuits."[82] In other words, Congress created this mess, and only Congress can fix it.

Sound forest management and wildfire prevention practices are neglected in the quest to fight wildfires through federal suppression policies. Although federal agencies struggle to manage both wildfire suppression and forest management, *they fail to advocate effectively for funding* and revert to status quo practices exacerbating poor management on all fronts. A telltale sign: internal employee satisfaction surveys at each federal agency rank leadership as weak, with the FS second to last of all the federal agencies surveyed.[83]

Unfortunately, the wildfire suppression and forest management practices adopted in the United States by government agencies remain counterproductive and increase the number and intensity of wildfires. Lack of forest management practices to clear dead trees and undergrowth, WUI, droughts, warming temperatures, and invasive, destructive insects only worsen and add to wildfire risk.

LEGISLATIVE OVERSIGHT

The federal government tried and failed to update wildfire management policies and funding multiple times. The only exceptions are the Wildfire Management Technology Act (2019), sponsored by Senator Maria Cantwell (WA), which requires firefighters assigned to wildfires to be equipped with GPS locators and drones carrying infrared sensors to map large wildfires in real-time, and the recent Infrastructure Bill

(2021) described later in this section.[84] Sadly, these bills are Band-Aids that fall far short of addressing our forests' predicaments.

In 2017 the USDA introduced the Forest Products Modernization (FPM) Strategic Framework for Long-Term Action: 2019-2023 to improve FS's management of forests. The FPM focuses on four areas: employee recruitment, retention, training, and education strategies; increasing efficiency through innovative technology; improving business practices; and updating and modernizing their policies. While the framework recognizes the importance of conducting restoration treatment to build more resilient forests, most of the FPM focus is on enhancing its timber management programs through improved technology and increased timber sales.[85]

In 2019 Senator (and later vice president) Kamala Harris (CA) introduced the Wildfire Defense Act to establish a wildfire defense grant program. It was referred to the Committee on Homeland Security and Governmental Affairs but did not receive a vote in Congress.[86]

In 2020 Senator Ron Wyden (OR) introduced a wildfire prevention bill (S. 4625) to the Senate in the 116th Congress. The National Prescribed Fire Act of 2020 would provide much-needed funding ($300 million) to increase the number of prescribed burns on federal, state, and private lands, including incentives for federal, state, and local agencies to conduct prescribed burns on areas larger than 50,000 acres. Unfortunately, like the above-mentioned Harris bill, this one, too, went nowhere in Congress and was reintroduced as S.1734.[87,88]

In 2020, and again in 2021, Senator Mitt Romney (UT) announced he was establishing a national commission to "study and recommend fire mitigation policies to Congress, including forest management tactics and federal spending." The Wildland Fire Mitigation & Management Commission Act of 2020 brings together federal and nonfederal stakeholders to review national wildlife policy and make specific policy recommendations for fire mitigation, management, and rehabilitation of forests and grasslands. In addition, the commission would develop a report that inventories firefighting aircraft and aircraft parts but not mechanized equipment. This proposal was incorporated into the recently signed (November 2021) Infrastructure Investment and Jobs Act.[89,91]

The Emergency Wildfire and Public Safety Act was introduced in 2020 in the Senate by Senators Dianne Feinstein (CA) and Steve Daines (MT) and by California Representatives Jimmy Panetta and Doug LaMalfa in the House. The bill would implement wildfire prevention projects, fund wildfire detection and response technology, and provide retrofitting assistance to make homes and businesses more resilient to wildfires. The Senate heard the Committee on Energy and Natural Resources bill, with no further action. After being introduced in the House, it was referred to several subcommittees, with no further action.[92-95]

More recently, the CARR Act, introduced in 2021 by Representative Doug LaMalfa (CA), would exempt federal wildfire mitigation activities conducted within three hundred feet of a road from environmental review laws. The bill was referred to several committees and subcommittees for review.[96]

The 21st Century Conservation Corps Act, introduced in 2021 by Representative Joe Neguse (CO), would provide federal land management and conservation funding. This act includes a provision to use the Reforestation Trust Fund to reforest certain federal lands. The act was referred to the Committee on Appropriations.[97]

In 2021 Senators Debbie Stabenow (MI) and Rob Portman (OH) and Representatives Jimmy Panetta (CA), and Mike Simpson (ID) reintroduced legislation to address this reforestation backlog. The Repairing Existing Public Land by Adding Necessary Trees Act (REPLANT) proposes eliminating the Reforestation Trust Fund cap and increasing the amount to $123 million annually. It also will reforest 4.1 million acres by planting 1.2 billion trees on national forests and create 49,000 jobs within the next ten years.[98,99]

Finally, the Infrastructure Investment and Jobs Act, enacted on November 15, 2021, contains initial, but insufficient funding from 2022 to 2026 for certain forest management activities. Its key forest management initiatives outlined in the legislation include:

- $500 million for mechanical thinning and timber harvesting, requiring each federal wildland firefighter to spend a minimum of 800 hours per year working on reducing hazardous fuel loads.

- $500 million for community wildfire defense grants.

- $500 million for prescribed burns.

- $500 million to study and develop fuel breaks.

- $800 million for ecosystem restoration, watershed protection programs, including financial assistance programs to purchase and process ecosystem byproducts.

- $200 million for the FS postwildfire area rehabilitation program.

- $250 million to restore, develop, and decommission trails and roadways to reflect long-term resilience.

- $20 million for the National Oceanic and Atmospheric Administration to enhance wildfire prediction and geostationary satellites to detect wildfire starts.

- $500 million to fund GNA and Shared Stewardship agreements.

- $200 million to fund detection, eradication, and "more research" on invasive species.[100]

This bill covers a five-year funding period instead of annual appropriations so agencies can improve long-range planning. While $5.5 billion was appropriated, regrettably, the bill does not address the substantial need for more mechanized equipment—specifically, aerial fighting aircraft and bulldozers to suppress wildfires. Also unaddressed: more year-round on-the-ground personnel to perform deferred maintenance of the nation's forests. On a promising note, the bill encourages public-private partnerships to fund GNA and shared stewardship agreements.

ENVIRONMENTAL LEGISLATION REQUIRING REASSESSMENT AND AMENDMENT

The National Environmental Protection Act of 1970 (NEPA) requires all federal executive agencies to prepare environmental assessments and impact statements for projects or activities impacting land, floral, mineral, or air use. These reports must include potential environmental effects of proposed federal agency actions. Further, Congress recognizes that *each person* is responsible for preserving and enhancing the environment as trustees for succeeding generations. NEPA also imposes criminal penalties for non-compliance.[101]

The administrative burden to prepare and successfully navigate federal and state bureaucracies is unreasonable and irrational. Often, forest management/stewardship projects require 800-page filings delaying timely execution of well-conceived and managed projects—sometimes by years. Parties opposed to the proposed actions use NEPA to petition courts to delay prescribed burns, sustainable timber harvesting, ecosystem enhancements, and other forest management activities to the detriment of the majority view. Fortunately, the Infrastructure Investment and Jobs Act enacted in 2021 established categorical exclusions for certain forest management activities.

Likewise, well-intentioned, environmentally conscious individuals hold key decision-making positions in government bureaucracies. Any bias creates a significant pendulum shift from a balanced approach to forest management to an overly sensitive environmental view of policy and practice. The "let burn" practices of wildfire and greenhouse gas management, causing the massive impact of wildfire smoke on every citizen's daily activity and health, is only one example of this pendulum shift.

Moreover, a rush to legislate is not wise. One argument against wildfire and forest management practices centers around the resulting destruction of wildlife habitats. An example is the spotted owl found in mature, older trees in western U.S. forests and listed as a threatened species since 1990.[102] Removing trees and letting naturally occurring wildfires burn are viewed by some as threats to the spotted owl population's continued existence. Some studies project the increasing intensity of wildfires and climate change could further jeopardize different subspecies of spotted owls.[103] The Northwest Forest Plan, enacted in 1994, reduced logging on 9.5 million acres of old-growth forests on federal land in part to save the spotted owl's habitats.[104]

A 2018 comprehensive review of wildfires and the spotted owl population concluded that wildfires are not a threat to the species and, in fact, may benefit the foraging habitat selection, recruitment, and even their reproduction. The spotted owl population declined 75 percent since 1994, not due to logging or wildfires, but due to its predator, the barred owl. It is not unusual for other animals and plants to thrive in areas burned by wildfires and sustain themselves for thousands of years.[105]

CASE STUDY—THE CARR FIRE[106]

On July 23, 2018, sparks from a trailer tire's steel rim ignited three separate grass and shrub brush fires northwest of Redding, California. While the brush fires were unpredictable accidents, what happened next was not.

Jumping from the roadside into the Whiskeytown National Recreation District, the fire rapidly increased in size and velocity, burning for 39 days and destroying more than 229,000 acres with an estimated loss of $1 billion. More than 1,000 homes were lost. The fire claimed the lives of eight people including four first responders.

The subsequent investigation showed that local, state, and federal personnel predicted and worried about such an event for years. Every level of government understood the dangers and took insufficient steps that might have prevented the catastrophe.

Roadside brush—the ignition source—was supposed to be co-managed by Caltrans and the NPS. Officials reportedly appealed to local transportation personnel where Route 299 took over local authority.

In 2016 Caltrans cleared roadside brush on its side of Route 299. Caltrans also planned to remove a four-foot to thirty-foot fire buffer strip from the pavement on the NPS side but ran afoul of NEPA's protection of natural resources/scenic values rules. Rather than hold a public hearing or file lengthy EPA environmental impact studies, alternative plans by NPS to undertake controlled burns were thwarted by local air pollution regulations. Redding residents and elected officials did not enforce WUI development regulations. Thin-from-below forest brush clearing plans were not sufficient due to budget constraints. It was estimated that $3.5 million per year used for thin-from-below practices could have prevented much of the Carr Fire destruction. The cost-benefit tradeoff is obvious.

Local WUI housing regulations have yet to be strengthened by local housing agencies, and despite the increase in 2021/2022 forest management budgets by NPS and California, the Whiskeytown National Recreation District budget still lacks the funding needed to reforest their destroyed terrain. Is the lack of better WUI regulations and dearth of funding for forest thinning and clearing setting up subsequent disastrous wildfires?

UNSUSTAINABLE POLICIES

Decades of inadequate wildfire and forest management practices have created vulnerable, unhealthy, and dangerous forests. These practices are unsustainable if the goal is to decrease destructive wildfires and restore our forests. This chapter casts a bright light on the current dysfunctional management practices of agencies responsible for forest management and demands a call for action to change. Furthermore, forest refurbishment requires financial resources currently beyond the scope of federal forest management officials or a bitterly divided Congress.

THE KEY TAKEAWAYS FROM THIS CHAPTER ARE:

- Investing in forest restoration and wildfire suppression will improve the health and resiliency of our national forests.

- Federal agencies struggle to conduct sustainable forest clearing, restoration, and revitalization due to lack of leadership and financial resources, bureaucratic inaction, and legal roadblocks.

- Inadequate WUI regulations over location, fire-resistant structural design, and enforcement increase life and property risk.

- Little, if any, year-round in the forest training exists, which creates an opportunity to for a more productive ground management to reduce forest floor fuels.

- To date, shared stewardship agreements containing reflective forest management practices represent just an idea and not meaningful actions to protect our forests.

- There is a severe lack of congressional action and oversight to provide needed financial assistance and oversight to federal forest managing agencies.

- Who will hold the USFS and DOI responsible for the funds committed and the promises made under the Infrastructure and Jobs Act?

- Private-public partnerships effectively manage precisely focused areas and projects and are now encouraged by recent infrastructure legislation.

- Native American forest stewardship is highly effective and deeply embedded in many tribal cultures.

- It is too early to determine whether recently enacted categorical exclusions to NEPA will allow well-conceived forest restoration projects and prescribed burns to proceed unimpeded by bureaucratic red tape. Historically, the real and perceived threat of litigation has had a chilling effect on agency forest management efforts.

- Over decades, federal and state inaction has created a substantial funding shortfall to restore forests to a healthy, sustainable condition.

CHAPTER 10

FOREST CARBON STORAGE, CARBON EMISSIONS, AND CARBON OFFSETS

Humanity is waging war on nature.... Nature always strikes back—and it is already doing so with growing force and fury.... Every country, city, financial institution, and company should adopt plans for transitioning to net zero emissions by 2050 ... (and take) decisive action now.

—Antonio Guterres, Secretary-General of United Nations[1]

This chapter describes the role of carbon and its impact on forests, forest management, and wildfire suppression. Practices encouraged by the United Nations Climate Framework have yielded positive results and, unfortunately, unintended negative consequences to our nation's forests from carbon emissions and offsets.

FORESTS CONTRIBUTION TO STORING CARBON

Carbon supports all life through its exchange between living things and the environment. Trees and plants absorb atmospheric carbon dioxide (CO_2) and release oxygen during photosynthesis, which moves carbon to their stems, trunks, roots, and leaves. When leaves or trees fall and decompose, or flora dies, the carbon stored in them releases and transfers back to the atmosphere or into the soil. Natural ecosystems can store considerable amounts of carbon and carbon "sinks," a term for critical systems of stocks or pools of global carbon.

These are Earth's five central carbon sink systems: (1) continental crusts and upper crusts of sedimentary rock formed over millions of years; (2) oceanic inorganic dissolved carbon; (3) atmospheric greenhouse gas (GHG); (4) soils, vegetation, and forests; (5) harvested wood products.

The diagram below (see fig. 1) depicts the global carbon cycle process from 1783 to 2019. During this 236-year span, forests and lands sequestered, on average, 34 percent of GHGs each year of a total of 9.5 gigatons of GHG emissions (a gigaton equals one billion metric tons or nearly 2.2046 trillion pounds), while oceans sequester another 26 percent. The remaining CO_2 circulates in the atmosphere. Together forests and oceans form a natural buffer against climate change, sequestering about 60 percent of GHGs each year.

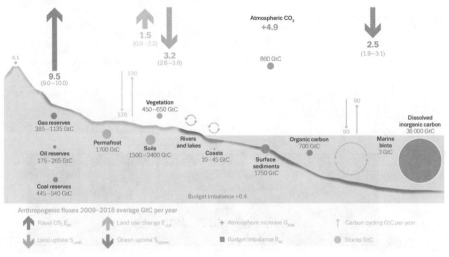

Figure 1. The global carbon cycle. (Figure by Copernicus Publications, December 4, 2019, https://essd.copernicus.org/articles/11/1783/2019/.) [Photo credit: The global carbon cycle from Global Carbon Budget 2019. This work is distributed under the Creative Commons Attribution 4.0 License.]

Forest carbon storage varies by region, climate, and tree stock. Warm tropical areas tend to store more carbon above ground unlike cool areas of boreal forests, which store enormous quantities of carbon below ground. *Flux* is the term that describes carbon entering or leaving soils, vegetation, and forests. The average rate at which carbon flows through these stocks—*carbon turnover*—provides the location where CO_2 might be released into the atmosphere from ecosystems across the globe.

Globally, 123 million gigatons are stored in carbon sinks, ocean sinks, continental crusts, and sedimentary rock sinks, accounting for 99 percent of total carbon storage. While forests sinks represent only a small fraction of this total (400 gigatons), they play a crucial role because next to oceans their cycle time to sequester CO_2 is shorter than other carbon sinks.[2]

The graph below (see fig. 2) depicts the most productive carbon stores by forest ecosystem type.

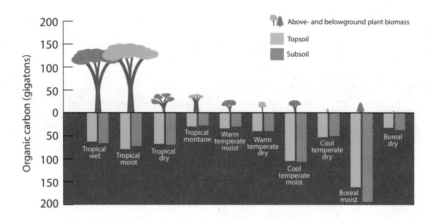

Figure 2. Carbon stored in global ecosystems 1990 to 2007. (Figure by USDA Forest Service, June 2017, https://www.fs.usda.gov/ccrc/topics/global-carbon.)

A significant amount of forest carbon is stored in woody biomass (roots, trunks, and branches). Another portion eventually becomes organic matter as forest floor litter decomposes into the soil. Increased tree growth rates increase forest biomass just as net carbon storage can result from reforestation and management growth.

The USDA Forest Service (FS) performs annual surveys and assessments of the nation's forests, data that includes the health and conditions of forests and the amount of carbon stored in forest biomass.[3] The United States Environmental Protection Agency (EPA) also analyzes inventory of greenhouse gas emissions and carbon sinks through its Greenhouse Gas Reporting Program. This comprehensive public report is provided to Congress and the United Nations (UN) as required by the Kyoto Protocol. The UN also summarizes global surveys submitted by member nations, also required by the Kyoto and Paris climate agreements. These reports provide annual analysis of carbon sink and carbon emission data in addition to carbon mass stored with gigatons as a constant measure.

GREENHOUSE GAS EMISSIONS

Greenhouse gas concentrations trap heat in the atmosphere. These gases include carbon dioxide (CO_2), methane (CH_4), nitrous oxide (N_2O), and fluorinated gases (hydrofluorocarbons, perfluorocarbons, sulfur hexafluoride, and nitrogen trifluoride). According to the EPA (see fig.3), "[E]ach of these gases can remain in the atmosphere for different amounts of time, ranging from a few years to thousands of years. All these gases remain in the atmosphere long enough to become well mixed, meaning that the amount measured in the atmosphere is roughly the same all over the world, regardless of the source of the emissions."[4]

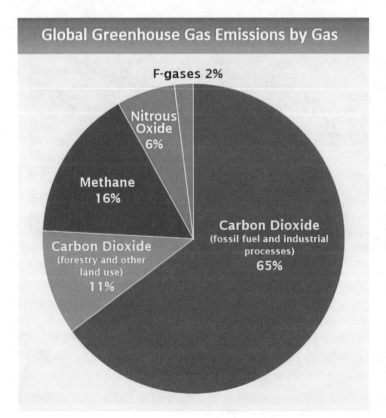

Figure 3. GHG emissions by gas. (Figure by Intergovernmental Panel on Climate Change, 2014, https://www.ipcc.ch/report/ar5/wg3/.)

GHG EMISSIONS BY GAS

The United States, China, India, the Russian Federation, and the European Union are responsible for most of the annual GHG concentration increases—64 percent of the global total.[5] The international economic sector contributors to GHGs include electricity and heat production; agriculture, forestry, and land use; transportation; and industry, or about 85 percent of annual GHG emissions.[6]

GHGs are measured in parts per million (ppm), parts per billion, and even parts per trillion. One part of GHG per million is equivalent to one drop of water diluted into thirteen gallons of liquid.[7] Another frequently used measurement is metric tons, which equal 2,205 pounds of GHG according to the EPA. A final conversion factor, one ppm, equals 7.8 billion tons of CO_2 in the atmosphere. (see fig. 4)

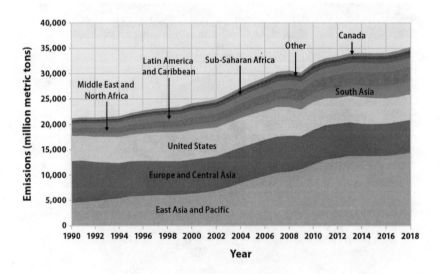

Figure 4. Greenhouse gas emissions, 1990–2018. (Figure by United States Environmental Protection Agency, April 2021, https://www.epa.gov/climate-indicators/climate-change-indicators-global-greenhouse-gas-emissions.)

CARBON DIOXIDE (CO_2)

As depicted in figure 5, carbon dioxide (CO_2) contributes more than 75 percent to global GHG each year. The National Oceanic and Atmospheric Administration (NOAA) reports the international average level of CO_2 in 2020 was 412.5 ppm, a 2.6 ppm increase above 2019. It was the sixth-largest annual increase since 1980 when NOAA's data estimated the average CO_2 atmospheric concentration of 338.9 ppm.[8] But the average rise of CO_2 drops to 0.55 percent per year during the same forty-year period.

> **Authors' Note**
> Global population growth from 1980 to 2020 grew 74.9 percent, from 4.458 billion to 7.795 billion, an average of 0.044 percent per year—the same growth rate as GHGs. In the meantime, Mother Nature continues her cycle of capturing from 55 to 60 percent of CO_2 emitted each year (see fig. 5).[9]

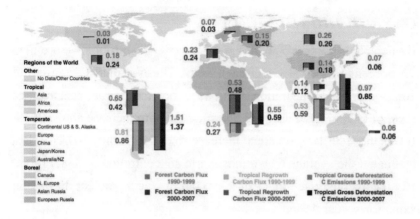

Figure 5. The global carbon cycle by continent. Colored bars in the downward-facing direction represent carbon sinks, whereas bars in the upward-facing order represent carbon sources (emissions). Light and dark purple = global established forests (boreal, temperate, and intact tropical forests); light and dark green = tropical regrowth forests after anthropogenic disturbances; and light and deep brown = tropical gross deforestation emissions. Note these are figures for carbon, not carbon dioxide. (Figure by the United Nations Forum on Forests, March 2019, https://www.un.org/esa/forests/wp-content/uploads/2019/03/UNFF14-BkgdStudy-SDG13-March2019.pdf.)

UNITED STATES FOREST CARBON DATA

U.S. forests stored 58.7 billion metric tons of carbon in 2020 (see fig. 6). Most forest carbon (95 percent) was stored in the forest ecosystem pools. The remainder was stored in the product pool—for example, harvested wood products. The most extensive collector of carbon was forest soils with 53 percent of total forest carbon in 2020. The next largest pool was above-ground biomass, which contained 26 percent of the total. Each of the other pools stored less than six percent of the total carbon.

Since 1990 in the United States, forest carbon stocks have increased 10 percent on a cumulative basis. Most forest pools gained more carbon as of 2020 compared to the Kyoto baseline year of 1990. The exceptions are the litter and soil pools, which each store around the same amount of carbon annually. Actually, forest carbon stocks are increasing annually, meaning U.S. forests are a net carbon sink, absorbing more carbon from the atmosphere than they release.

About one-third of the United States is forested. As described in chapter 8, forested areas vary by location, climate, vegetation type, disturbance histories, and other factors. Because of this variation, amounts of carbon vary across the different forest pools (see fig. 6). Accordingly, the amount of carbon within a particular area or carbon density also changes.

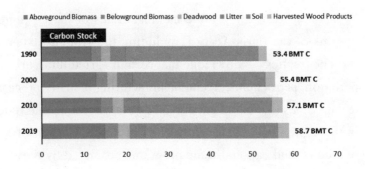

Figure 6. U.S. Forest carbon stocks by pool (billion metric tons [BMT] of carbon [C]). (Figure by Congressional Research Service, May 5, 2020, https://crsreports.congress.gov/product/pdf/R/R46312.)

THE KYOTO PROTOCOL, PARIS ACCORD, CARBON OFFSETS, AND CAP-AND-TRADE

Adopted in 2005, the Kyoto Protocol requires parties in the developed world to limit GHG emissions relative to their 1990 emission levels. The UN-sponsored treaty, ratified by 192 countries, promotes the use of "carbon offsets." Carbon offsets are derived from projects either committing to increase carbon sinks or projects dedicated to avoiding future GHG emissions. A carbon offset is a "claim" on that project, which can be located anywhere on the planet. A carbon offset or credit compensates an entity acquiring them for GHG emissions produced, which counts as a reduction of their carbon footprint. The Kyoto treaty's thesis and the subsequent Paris Accord agreement affirm carbon offsets and credits to be applied against an equivalent carbon footprint. The 2015 Paris Agreement does not mention carbon markets or carbon offsets. Instead of offsets, Article 6 of the Paris Agreement refers to "internationally transferred mitigation outcomes." Government negotiations are not yet complete at the United Nations Framework Convention on Climate Change on the rules for a new carbon trading mechanism under Article 6.

By January 2021, all but five member states of the UN agreed to the terms of the Paris Accord, including the United States and Canada. The treaties acknowledge that certain agricultural, utilities, transportation, petrochemical, and manufacturing industry processes lack technologies that could encourage capital investment to reduce or eliminate the GHGs these industries produce. Carbon offsets, carbon credits, and cap-and-trade programs promoted by these two United Nations Climate Change Frameworks allow carbon-emitting sectors to reduce their reported emissions by applying carbon offsets and credits they acquire.

Carbon offsets and carbon credits. A *carbon offset* is defined as an instrument representing the reduction, avoidance, or sequestration of one metric ton of CO_2 or greenhouse gas equivalent. It creates carbon credit generation from a project with clear boundaries, title, project documents, and a verification plan. Carbon offsets are traded voluntarily through a marketplace to generate reductions in the carbon emission footprint from outside a company's physical premises. Projects such as planting trees, preserving forests, developing wind farms, and investing in renewal energy are common carbon offset projects and represent actions to reduce atmospheric CO_2. Although voluntary, offsets may be mandated by a government agency.

A *carbon credit* is distinctly different from a carbon offset. A carbon credit represents ownership of one metric ton of carbon dioxide equivalent that can be traded, sold, or retired. Companies regulated under a cap-and-trade system have an allowance of credits they can use toward their cap. When fewer emissions (credits) are used than allocated, the company can trade, sell, or hold the credit. If sold, this allowance of emissions transfers to another company. When use exceeds allocation, regulated companies must purchase a credit to comply with cap-and-trade regulations. Like an offset, a credit becomes tradable only when emissions are reduced. If carbon reductions acquired from carbon credits or offsets are equivalent to the total carbon footprint of an activity, then the action is described as "carbon neutral."

Cap-and-trade. Cap-and-trade is a government regulatory program designed to limit or cap total level of emissions of certain chemicals, especially carbon dioxide. Regulatory caps established are industry specific. Currently, cap-and-trade regimes exist in China, the Republic of Korea, the European Union, California, and Canada's Quebec Province.

A government agency issues a limited number of annual permits permitting companies in a defined industry, such as petrochemical refiners, to emit a certain amount of carbon dioxide, which becomes the cap on emissions. Companies are taxed if their emission exceed the stated level in their permits. Companies that reduce their emissions below their cap can sell or trade to other companies, enabling the acquiring entity to reduce its reported emissions. In addition, companies can accumulate unused permits in years when they emit less CO_2 than established by that year's cap. These unused credits can be carried forward to offset future year CO_2 emissions.

Each year the government lowers the number of permits, thereby reducing the total emissions cap, making the permits more expensive. The reason is it provides an incentive for companies to invest in clean technology as it becomes less expensive than permits.

CARBON-OFFSET MARKETPLACE

The Kyoto Protocol and Paris Accord endorsed and encouraged the establishment of a global marketplace for the purchase, sale and trade of forest carbon offsets and carbon credits. It marked an historic first: a process allowing nations, consumers, and businesses to voluntarily buy, sell, and trade carbon offsets and credits derived from forest carbon projects to compensate for their GHG emissions.

A tradable certificate or permit represents the claim on the underlying carbon project. Amounts are measured in tons of carbon dioxide-equivalent CO2e. One ton of carbon offset derived from a project represents the reduction of one ton of carbon dioxide or its equivalent in other GHGs. Market prices for carbon offsets and credits vary based on the quality of the underlying project, the validation process used to quantify the volume of the carbon, contract length, and plans for the reduction in CO_2e emissions among other factors.[10]

Large purchasers of carbon offsets include companies aspiring to be carbon neutral (major airlines, oil and gas producers, technology companies, and major manufacturers). Nations with positive carbon footprints can trade carbon offsets to net carbon-emitting countries, allowing the emitters to comply with the United Nations Climate Framework targets.

Although a few international quality standards exist, the voluntary market for carbon offsets is unregulated. An industry task force developed comprehensive recommendations to improve transparency, integrity, verifiability, and certification practices.[11] Recommendations from the task force indicate this evolving marketplace is still in a very nascent stage.

How do forest carbon offsets work? GHG emitters can purchase them to avoid or offset their emissions occurring elsewhere. According to the Climate Trust, there are generally three types of forest carbon offsets:[12]

- Afforestation and reforestation: Sequestered carbon and offsets generated through the creation or re-establishment of forests.

- Avoided conversion: Forests with a high likelihood of tree and carbon loss (usually from conversion to agriculture or development) commit to retaining the forest. Carbon dioxide emissions avoided through this conservation effort yield offsets.

- Improved forest management: Better, sustainable forest management increases carbon in the forest and durable harvested wood products. Improved forest management projects may include:

  » Increasing overall age of the forest by increasing rotation ages.

  » Increasing forest productivity by thinning diseased or suppressed trees or managing brush and other competing vegetation.

  » Improving harvest practices.

  » Maintaining stocks at an elevated level.

  » Increasing the level of carbon stocks and sinks.

Carbon offset programs anticipate risks like wildfires and provide an appropriate cushion or insurance by setting apart a portion of the total purchased offset credits in anticipation of potential disasters. The market recognizes this insurance requires calibration to reflect current and future climate risk.

Several independent registries record and retire projects offered for carbon offsets, including the UN's Clean Development Mechanism, Verified Carbon Standard, Gold Standard, American Carbon Registry, Climate Action Reserve, Plan-Vivo, and Architect for REDD+ Transactions. REDD+ stands for "Reducing emissions from deforestation and forest degradation and the role of conservation, sustainable management of forests and enhancement of forest carbon stocks."[13] None of these registries is currently regulated, meaning each registry may have different requirements and audit standards for the carbon offset project. Should changes occur during the life of an offset project, not all registries require notification. Neither is there an enforcement arm to ensure project integrity reporting throughout the project's life. Carbon offset forestry contracts represent permanent emission reductions based on terms that can run from 40 years or 100 years, reflecting the long-term investment required to mimic the full life-cycle development of forests.

- According to Forest Trends' Ecosystem Marketplace, through 2019, the cumulative market volume of voluntary carbon offset projects topped 1.3 billion tons, with a value exceeding $5.5 billion.[14] McKinsey estimates this market could grow to $50 billion by 2030.[15] This fast-growing market reflects global demand for carbon offsets by an increasingly significant number of CO_2 emitters.

- Established exchanges or marketplaces allow participants to buy and sell individual carbon offset projects in secondary markets. The largest include European Climate Exchange, the NASDAQ OMX Commodities Europe Exchange, and the European Energy Exchange.

REGULATORY AND STRUCTURAL SHORTCOMINGS OF CARBON OFFSET AND CAP-AND-TRADE MARKETPLACES

Some observers call these evolving marketplaces "the Wild, Wild West" because of glaring regulatory shortcomings. Recommendations developed by the industry taskforce provide a daunting list.[16] These include:

- Most countries currently do not have a legal framework in place to explicitly allocate carbon rights. Instead, carbon rights must be inferred from existing law considering the specific project context.

- Compliance with obligations contained in the carbon offset contract lack meaningful enforcement mechanisms. Should a buyer default on the ongoing forest management of the acreage covering a carbon credit permit, enforcement is strictly voluntary with little or no regulatory oversight.

- To be effective, a carbon offset must be incremental; that is, the project must reduce GHG emissions at a greater rate than would have occurred in the absence of the offset. Thus, the carbon benefits must be accretive relative to outcomes absent the carbon offset.

- Permanence of the emission-reduction project requires examination. A tree planted to offset carbon emissions should not be removed in the future. In addition, carbon offset projects can create leakage where a project's impact unintentionally increases emissions elsewhere, e.g., when deforestation is relocated rather than avoided.

- A carbon offset or carbon credit is not a formal secured claim on any portion of the forest carbon project.

- Land use rights vary by federal, tribal, state, and local regulations, meaning the formal contracting process can be challenging and expensive.

- Insurance coverage for acreage destroyed by wildfires covered by a carbon offset may not be adequate to compensate for the risk of loss to the offset buyer.

- One issue in establishing a cap-and-trade policy is the amount of the government-imposed cap on emissions producers. A significantly high cap may lead to even higher emissions, while a notably low cap might be viewed as a burden on the industry and a cost to be passed on to consumers.

- Unused emission credits under cap-and-trade regimes can be "banked" and carried forward to offset future emission caps because the initial cap established by government agency was too liberal.

- Environmental activists argue that a cap-and-trade program will prolong the active life of polluting facilities by allowing companies to delay action for years until it becomes economically infeasible.

- Under cap-and-trade platforms, industry caps are established by regulators allowing the more severely emitting companies to defer emission cutback investments and practices.

- Offsets on public lands may be highly problematic because they can be a deterrent to active forest management.

- Cheap carbon offsets can be used to avoid the challenging work of actually cutting emissions. "The practice is so common that critics often describe the certificates as 'papal indulgences,' reminiscent of the way Catholics in the Middle Ages made payments to the Church to eliminate the stain of sinful deeds," reported a Bloomberg article. "Should companies be allowed to balance their carbon ledgers by purchasing offsets without first exhausting all other options to clean up their business?"[17]

CASE STUDY—COLVILLE RESERVATION TRIBES AND BP

In 2016 BP (a $185 billion global petrochemical company) paid a Washington-based confederation of indigenous tribes $19 million to secure a carbon offset for a half-million-acre forest maintenance and tribal economic development.[18] The offset contract was for an unspecified term. It required the tribes to maintain carbon levels through sustainable forest management practices.

In 2019 the nearby William Flats Forest caught fire. The Bureau of Indian Affairs made an initial "let burn" decision on the fire, which later expanded, destroying 44,500 acres, including the forest under the restoration contract with BP. Since the forest was destroyed, carbon sequestration offsets were no longer available under the restoration offset project. BP demanded repayment of the $19 million in funds advanced for the project from the Colville indigenous tribes. Fortunately, the Confederation of Colville tribes had wildfire insurance for its forest that fully paid BP's claim. BP did lose its forestry carbon offset rights.

Imagine the consequences had the wise financial managers of the Colville confederation not secured wildfire insurance. Litigation most certainly would have ensued negatively impacting the financial condition of the tribe.

This case study demonstrates one of the many risks of purchasing forest carbon offset. Forestry carbon offsets lack long-term reliability, putting into question their value as a tool in the climate change fight.

The United Nations Climate Change Framework encouraged trading carbon credits between countries, particularly for accretive carbon projects in undeveloped nations. The controversial call to action caused many UN member states to examine national policies related to national forests and the advisability of granting carbon offsets using sovereign forest assets.

REDD+ is a framework that aims to curtail climate change by curbing destruction of forests. The "+" signifies the role of conservation, sustainable management of forests, and enhancement of forest carbon stocks. REDD+ is the framework through which countries, the private sector, multilateral funds, and others pays countries *not* to cut down their forests. This practice can take the form of direct payments or an exchange for carbon credits. Currently, the United States does not allow its national forests to participate in any carbon offset or carbon credit markets. Yet, private landowners, American Indigenous nations, and certain states participated in the carbon offsets and carbon credits marketplace with their forests. As a result, these entities are paid to *not* thin particular forests creating an immediate conflict with best forest management practices to curtail potential wildfire threats. The unintended consequence of the UN policy increases wildfire risk absent strict guidelines. Unless carbon deforestation agreements avoid creating dense forests, more woody biomass surface cover, and higher fuel loads, wildfire risk may substantially increase.

The FS recently unveiled a study group to examine the future role of carbon credit transactions. This potential seismic paradigm shift could result in forest common credit sales to third parties solely because of the financial benefit that may ignore potential national security interests. Sovereign rights can become an issue as demonstrated in the following Peru-Switzerland case study.

CASE STUDY—SWITZERLAND AND PERU EXECUTE WORLD'S FIRST INTERNATIONAL CARBON OFFSET AGREEMENT[19]

In an agreement between Switzerland and Peru in October 2020, Switzerland agreed to fund an energy-efficient cooking stoves project in Peru called "Tuki Wasi." The project claims that reduced deforestation from more efficient cooking stoves will generate 100,000 carbon offsets by 2030 due to less forest wood use.

To fund its payments, Switzerland plans to impose taxes on fuel importers who would pass them on to Swiss motorists. Switzerland will allocate 100,000 carbon offsets to its country's CO_2 emitters so they can report a corresponding reduction in CO_2.

Even if this offsetting agreement results in reduced emissions in Peru (which is highly at risk), it will not reduce emissions globally. Every carbon offset generated in Peru will result in one ton of GHG emissions continuing in Switzerland.

Offsetting does not reduce emissions. The offsetting agreement between the two countries will help Switzerland meet its emissions targets, making it more difficult for Peru to do the same. The offsetting arrangement also creates a pertinacious incentive. The lower Peru sets its emission reduction targets, the more carbon offsets will be available for sale.

Each EPA's annual emissions inventory report exceeds 700 pages. It contains extensive details on GHG emissions, including GHG components and sources, size, composition of carbon sinks, and other data required by the UN.[20] In preparing this report, EPA collaborated with hundreds of experts representing more than a dozen U.S. government agencies, academic institutions, industry associations, consultants, and environmental organizations. The EPA also collects GHG emissions data from individual facilities and suppliers of certain fossil fuels and industrial gases through the Greenhouse Gas Reporting Program.

Figure 7 summarizes the latest EPA data on GHGs emitted by the United States into the atmosphere yearly from 1990 to 2019 (see fig. 7).[21] The United States emitted 6.6 billion metric tons of GHGs in 2019, a decrease of 864.7 million metric tons (mmts) from 2005 or a 12 percent reduction. From 1990 (the emission target required under the Kyoto Protocol) through 2019, GHG emissions only increased 115.6 mmts or 1.8 percent.

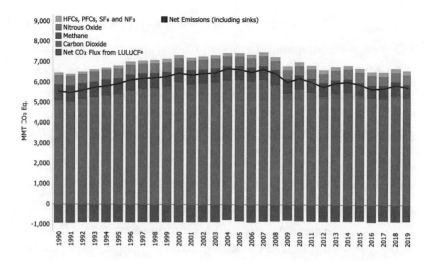

Figure 7. U.S. greenhouse gas emissions by gas. (Figure by United States Environmental Protection Agency, April 14, 2021, https://www.epa.gov/sites/default/files/2021-04/documents/us-ghg-inventory-2021-main-text.pdf?VersionId=wEy8wQuGrWS8Ef_hSLX-Hy1kYwKs4.ZaU.)

By comparison, during this same thirty-year period, the U.S. population grew from 248.7 million to 328.2 million people—a 32 percent increase.[22] On a per capita basis, U.S. GHGs declined 30 percent!

TECHNOLOGIES TO REMOVE CO_2 EMISSIONS

Policy makers, engineers, and scientists seeking to mitigate global warming proposed innovative technologies to sequester carbon. Technologies include geoengineering proposals called carbon capture and storage (CCS). In CCS processes, CO_2 is first separated from other gases contained in industrial emissions, then compressed and transported to a location isolated from the atmosphere for long-term storage. Suitable storage locations might include such geologic formations such as deep saline, depleted oil and gas reservoirs, or the deep ocean.

Although CCS typically refers to the capture of CO_2 at the source of emission before release into the atmosphere, it also may include techniques such as scrubbing towers and artificial trees to remove CO_2 from the surrounding air. So far, CCS is unproven and expensive. Despite billions of dollars invested, no CCS project has been delivered on time, budget, or agreed upon performance results.[23]

CO_2 EMISSIONS CREATED BY U.S. FOREST FIRES

A forest fire removes carbon from the forest releasing CO_2 into the atmosphere. The EPA, working with the FS's Inventory and Analysis annual report, reported the estimate of CO_2 emissions from forest fires. Europe's Copernicus Atmosphere Monitoring Service also monitors wildfire CO_2 emissions. The chart below summarizes both agencies' findings and includes prescribed burn and wildfire data (see fig. 8).

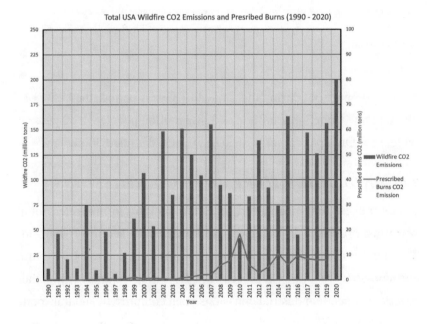

Figure 8. U.S. wildfire and prescribed burn CO2 emissions 1990 to 2020. All information from EPA 2016, 2018, and 2021 Inventory Reports, except totals in 2019 and 2020, which was reported by Europe's Copernicus Atmosphere Monitoring Service (2020 data through September 2020). (Figure by © Crowbar Research Insights LLC™. All rights reserved.)

As summarized in the above chart, CO_2 emissions from wildfires and prescribed burns impact GHGs each year. Harmful emissions trajectory increases yearly as more extensive and more intense wildfires consume our forests. Based on this data, on average, wildfires emitted 140 million metric tons of CO_2 in the last five years and more than 200 million metric tons in 2020. Future projections of more intense wildfires could make those 2020 emissions the norm in the next decade.

Note the increase in CO_2 emissions from prescribed burns. Chapter 8 described the critical role prescribed burns can play in forest management, but as seen in this data, prescribed burns come with a cost of higher CO_2 emission risk. Additionally, as described in chapters 1, 3, and 8, wildfire smoke emissions produce carcinogenic toxins, impact air quality of substantial portions of the population, force employers to ask employees to stay at home instead of their workplace and produce harmful GHGs.

What is the potential impact on our environment, emissions, and livelihood if wildfires are extinguished immediately instead of being allowed to burn for weeks and even months? We could enjoy the recreational experience provided by our forests, breathe clean air, and release fewer GHGs into the atmosphere. We might even achieve the 1990 Kyoto Protocol targets, thereby doing our part to reduce global warming.

THE KEY TAKEAWAYS FROM THIS CHAPTER ARE:

- Carbon sinks sequester vast quantities of CO_2, with oceans and forests taking up 60 percent of annual GHG emissions each year.

- Annual reporting of GHGs, carbon sinks, and wildfire CO_2 emissions provide ready access to the progress made in alignment with the United Nations Framework on Climate

Change established with the Kyoto Protocol and Paris Accord climate agreements.

- On a per-capita basis, GHGs decreased 30 percent in the United States since 1990.

- In 2019 U.S. GHG annual emissions increased only 1.8 percent from the 1990 Kyoto Protocol baseline target despite a 32 percent increase in the U.S. population.

- Putting out wildfires significantly reduces annual CO_2 emissions providing a viable source to meet Kyoto Protocol GHG targets for the United States.

- The regulatory and voluntary marketplaces where carbon offsets are traded creates substantial governance, land use, enforcement, and risk management challenges. Additionally, the rapid growth in these marketplaces may outstrip the ability to include meaningful safeguards to protect buyers and sellers from significant adverse consequences.

- Cap-and-trade offsets and carbon offsets may not reduce global CO_2 emissions, especially without robust verification processes, which are not currently in place.

- Although promising carbon sequestration technologies are under development, they require substantial capital to the market, thereby questioning their cost-versus-benefit outcomes.

- Carbon credit transactions involving U.S. forests require extensive public scrutiny and a go-slow approach to any such transactions.

CHAPTER 11

PUT THE FIRES OUT! ANALYSIS AND RECOMMENDATIONS

Now, this is not the end. It is not even the beginning of the end. But it is perhaps the end of the beginning.

—*Winston Churchill*[1]

Put the fires out! That's the first priority.

What more compelling evidence is needed to change the way we fight wildfires? How much more will we tolerate? Loss of life, carcinogenetic smoke inhaled by as much as two-thirds of the U.S. population, annual disruptions to work and home, decreased forest recreational use, annual damage costs as high as $300 billion and the environmental harm caused to our carbon stores by wildfires ought to be forceful enough evidence.

What's more disturbing? To our knowledge, no country other than the United States condones a "let burn" wildfire strategy. This practice defies responsible forest stewardship (other than for very remote, densely forested areas). Practically every country adopts "put the fire out first" practices. In contrast, the USDA Forest Service (FS) practices on federal lands—particularly in the western, most fire-prone states—inexplicitly vary. Stop the nonsense and just put the fires out!

Under current federal agencies' practices, wildfires now place 46 million residences in 70,000 communities at risk.[2] Two-thirds of the country face the threat of large, long-duration wildfires (see fig. 1). As the wildland-urban interface (WUI) expands due to expected population growth in the next twenty-five years, some experts predict a 50 percent increase in wildfire acreage consumed by 2050.[3]

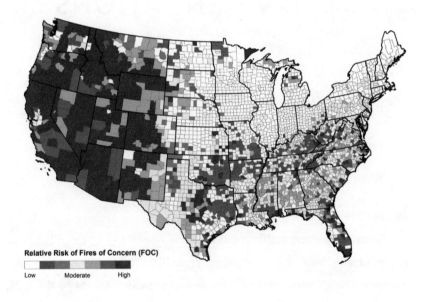

Figure 1. Areas in the United States at risk for large, long-duration wildfires. (Figure by the Science Analysis of the National Cohesive Wildland Fire Management Strategy, 2022, https://cohesivefire.nemac.org/option/10.) [Figure credit: NEMAC and the Eastern Forest Environmental Threat Assessment Center (EFETAC).]

Despite these ominous facts, the last update to federal interagency wildfire fighting policy was in 2009 (see chapter 2). It excludes any mention of prioritizing early wildfire extinguishment. Furthermore, the latest wildfire strategy issued by the FS in January 2022 relies heavily on forest thinning practices, such as mechanical removal of trees and prescribed burns. The strategy deemphasizes wildfire suppression. Establishing a clear strategy to prioritize putting out the fire compels each wildfire fighting agency to revise its mission statements. A put-the-fire-out-first strategy should be fundamental, along with determining expected response times, equipment requirements, desired forest conditions, and organizational structure.

Annual devastation from wildfires requires an immediate, laser-focused, and warlike response. Study after study shows aggressive wildfire initial response within the first few hours of ignition minimizes the likelihood of more devastating and intensive wildfires. Why invest millions of dollars annually in decision support analytics, predictive systems such as satellite imagery, remote weather accessible stations, and national alert warning systems only to adopt a policy to reduce wildfire suppression?

Are we foolish enough to believe, as stated in the latest wildfire crisis policy, that thinning two million acres of forest, an increase from the current rate of 800,000 acres, for each of the next ten years will control wildfire incidence or intensity? More than 100 million acres of national forests are at risk of severe wildfires and in need of active forest management (see fig. 2).[4] Agencies refuse to acknowledge that manpower for accomplishing this task is woefully in short supply—a lesson they failed to learn from the continuing labor shortage currently plaguing the FS. Announcing an increase of fuels and forest health treatments in western forests while maintaining and then reducing wildfire suppression costs isn't a Hobson's choice; it's foolhardy and irresponsible.[5]

The following chart shows that under the January 2022 U.S. Forest Service Wildfire Crisis Strategy, acres at risk of wildfire continue to grow rapidly despite announced plans to increase forest thinning to 20 million acres in the next ten years. At the proposed rate of thinning, wildfire risk increases rather than decreases as public statements from federal forest officials lead us to believe. Their plans simply do not even get ahead of insect damage, much less reduce unmanaged fuel loads mounting each year on the nation's forest floors.

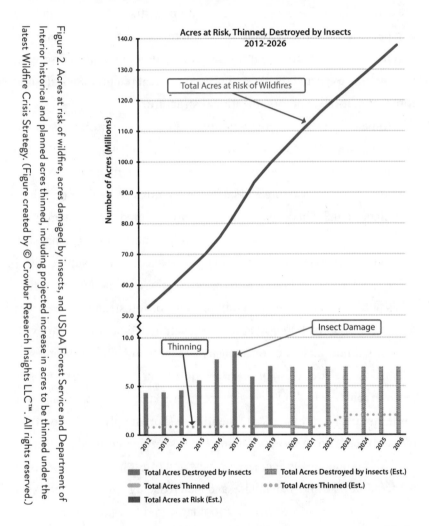

Figure 2. Acres at risk of wildfire, acres damaged by insects, and USDA Forest Service and Department of Interior historical and planned acres thinned, including projected increase in acres to be thinned under the latest Wildfire Crisis Strategy. (Figure created by © Crowbar Research Insights LLC™. All rights reserved.)

The rate of insect infestation projected in the next ten years likely will exceed the announced increase in forest thinning, thereby negating any significant reduction in the 100 million acres of under- and mismanaged forests.[6] Our assessment predicts the total undermanaged and mishandled acres will exceed 135 million by 2026 following implementation the newest thinning strategy. Increased forest thinning means hiring more workforce capacity in addition to more than 40,000 employees already responsible for forest management with little to no increased impact on wildfires. At the same time, it will add more overhead across the federal government.

The FS and Department of Interior (DOI) officials refuse to acknowledge these challenges. Forget a fist-pounding approach at congressional appropriation hearings demanding more financial resources to tackle the crisis. Instead, these bureaucracies choose a path lacking meaningful strategies to protect Americans and the forests making up our natural resources. Maintaining the status quo is shameful!

The American public deserves a more focused and professional federal agency to manage the wildfire and forest management crisis to protect its citizens, property, wildlife, and environment. Based on our research, we conclude the capabilities of the existing agencies, particularly at key leadership positions, require a rigorous independent evaluation. The scope of the U.S. forest management and wildfire suppression mission requires executives highly skilled in managing substantial organizations, not homegrown personnel with little or no experience other than in forestry. We also recognize the excellent work of the core middle-management government employees and recently recruited younger personnel. This group has a can-do, energetic passion.

But our findings suggest a significant restructuring of the current departments and agencies is required because the path chosen to date by the executive leadership continues to defy common sense in addition to business, economic, and basic management principles as wildfires are treated as an annual ritual instead of catastrophes.

How and when will this Potemkin façade cease?[7] Start by notifying our representatives and senators in Congress who refuse to protect citizens affected by wildfires that government bureaucracies and congressional committees are not doing their jobs.

Remarkably, the House Agriculture Committee and Senate Committee on Agriculture, Nutrition, and Forestry and their staffs are responsible, accountable for funding the various agencies protecting our forests and fighting wildfires. It's not too late to start asking the hard business questions. Any commonsense analysis demonstrates the agencies charged with forest maintenance for the past 117 years should get a grade of F for their inability to take forward-thinking actions. They get an A+ for conducting research, holding commission hearings that go nowhere, and establishing a bureaucratic organization so complex it defies logic. We summarized the growth of this bureaucracy in chapter 2 and reemphasize it here (see figs. 3 and 4).

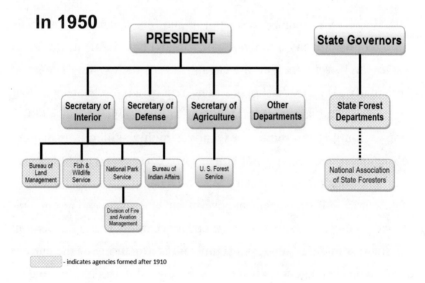

Figure 3. Organizational relationship of forest and wildfire management agencies in 1950. Congress authorized three new agencies and one new division between 1910 and 1950. States founded their forest departments starting in 1913. (Figure created by © Crowbar Research Insights LLC™. All rights reserved.)

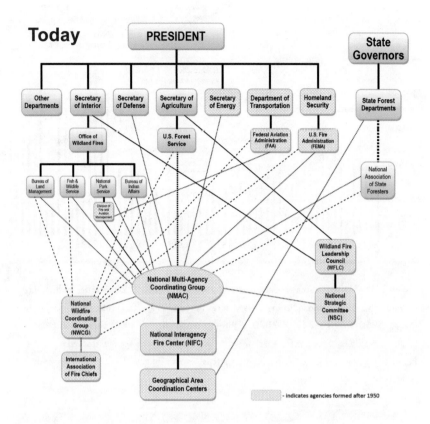

Figure 4. Organizational relationship of forest and wildfire management agencies in place in 2005 and continuing through today. In the fifty-five years since 1950, Congress authorized these new departments and divisions, agencies, councils, and programs. (Figure created by © Crowbar Research Insights LLC™. All rights reserved.)

Let's review critical metrics summarizing the performances of these departments, agencies, councils, and centers using figures developed for other chapters. Whether looking at the number of acres burned, number of structures destroyed, volume of wood harvested, or number of prescribed burns, the trend lines demonstrate poor results on each of these key performance indicators (see figs. 5 and 6).

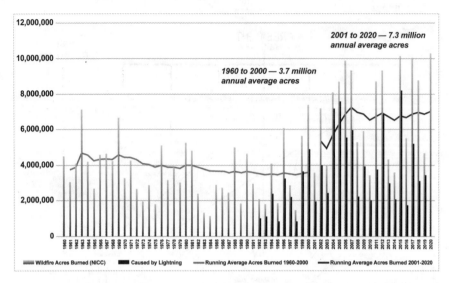

Figure 5. The United States acres consumed by wildfires 1960–2020. Annual wildfire acres burned 1960–2020, and lightning-caused acres burned 1992–2020 with a running yearly average 1960–2020 and 1992–2020. Lightning-caused fire reporting began in 2001. (Figure created by Crowbar Research Insights LLC™ with information extracted from the National Interagency Coordination Center, https://www.nifc.gov/, © Crowbar Research Insights LLC™. All rights reserved.)

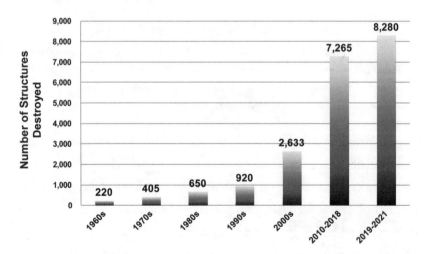

Figure 6. The number of structures destroyed by wildfires. The last three years eclipsed the combined losses covering forty years through 2009. (Figure created by Crowbar Research Insights LLC™ with information extracted from the National Interagency Fire Center, © Crowbar Research Insights LLC™. All rights reserved.)

More than 750 lives have been lost to wildfires from 2014 to 2021 and economic losses between $50 billion and $300 billion each year impact thousands more Americans. If these metrics aren't enough of an incentive for change, consider the environmental damage from 200 million metric tons of carbon dioxide (CO_2) released into and permanently impacting the atmosphere each year from wildfires along with the corresponding carcinogenic smoke affecting two-thirds of our country's citizens.

It's time to stop accepting these agencies' decades-long smokescreens created by the leaders responsible for wildfire and forest management. Congressional leaders should recognize they have two of the greatest research and investigative teams at their fingertips: the Congressional Research Service (CRS) and the U.S. Government Accountability Office (GAO). Why create yet another commission or congressional committee work group on wildfires when the end result is inaction on findings reported? Common sense dictates use of the CRS and GAO for help in understanding the facts about fighting wildfires, the leadership required, and the genuine issues that must be resolved to preserve our forests.

The record compels the agencies and legislature to change their behavior. Australia, Canada, China, Russia, and Spain spend more of their budgets on forest management than the United States. Congress has allowed undermanagement and underfunding our forests by establishing an unaccountable and unwieldy bureaucratic apparatus. It's time to reorganize the bureaucracy and make it responsive to the obvious and imminent national dangers. Incremental change hasn't worked. Major transformations are needed. Here are our recommendations:

CREATE A DEPARTMENT OF NATURAL RESOURCES

We resurrect past proposals to consolidate the various agencies involved with protecting our natural resources into one Department of

Natural Resources. Efficiencies identified in previous GAO reports will likely surpass $1 billion.[8] These savings could fund forest floor fuels removal projects.

It makes no sense to continue housing the FS within the U.S. Department of Agriculture (USDA). Take another look at the insanity of the current organizational structure outlined in figure 4 above. One obvious example: the National Oceanic and Atmospheric Administration (NOAA) should not report to the Department of Commerce.

Five separate studies under several presidents along with the CRS in 2008, the GAO in 2009, and various media op-ed pieces, in particular *The Washington Post* in 2020, recommend consolidating agencies. The only dissent comes from embedded, entrenched bureaucrats.

Stand up to the bureaucrats, Congress—it's your job! Establish a new Department of Natural Resources combining the FS, NOAA, Soil Conservation Service, and DOI under one management structure. The new organization could eliminate the apparent duplication created by redundant agencies, councils, and administrations. Terminate confusing unaccountable practices and install executive leadership with experience in managing large, complex organizations within and outside the forestry community. This wake-up call will establish a clear policy for all the nation's natural resources including logical forest management, and create an organization designed for action, accountability, and results.

ENACT LOCAL, STATE, AND FEDERAL LEGISLATION FOR THE WILDLAND URBAN INTERFACE (WUI)

As we documented earlier, property damages and subsequent loss of life in the past twenty years directly correlate to encroachment into the WUI. Population growth, demographics, higher living standards, and flight from urban centers contribute to home building in higher-risk wildfire areas. Urban flight will force smaller communities' policy makers to adjust master plans, infrastructures, and zoning practices. An article in the *New York Times* estimated 12.7 million more houses and 25 million more people living in WUI areas in 2010 than in 1990.[9] This trend only will increase over the next thirty years. For example, Colorado anticipates a 40 percent growth in residences by 2050.[10] Technology advances in video conferencing make lifestyle changes possible as more of the workforce can work from home and move closer to nature.

Once again, numerous reputable studies identify best-practice community fire adaptation criteria pertaining to building codes, hardened fire-resistant building materials, buffer zones surrounding structures, and defensible vegetation. But communities have been slow to adapt, increasing wildfire risk. Property insurers now substantially increase homeowners' premiums or deny coverage for those living in wildfire higher risk areas. Federal and state forest officials may continue to tout prescribed burns, but local air quality regulations restrict planned events.

We criticize state legislators for not doing the following:

- establishing local land-use policies and laws to require wildfire hazard mitigation;

- modifying land-use regulations to mandate fire-resistant materials use;

- expanding educational programs to inform WUI communities on how to create and improve fire-adapted communities; and

- committing funds to proactive, effective mitigation strategies, such as thinning and grazing (see figs.7 and 8).

Figure 7. Hundreds of sheep from Cuyama Lamb graze along the hillside in Southern California. (Photo by ABC News, November 13, 2021, https://abcnews.go.com/U.S./flocks-sheep-firefighting-solution-knew-needed/story?id=80800627.)

Figure 8. Healthy forests, healthy communities. (Photo by Healthy Forests, Healthy Communities, https://healthyforests.org/issues/.)

WUI communities require education on the benefits of maintaining healthy, sustainable forests and the risks when they are not. Recent studies demonstrate the better informed the residents, the less chance for human ignition-caused fires.[11]

State and local WUI communities should expand the use of innovative social media applications to increase public acceptance of fire and fuel treatment efforts. Social media platforms could inform and financially incentivize community behavior to remove and dispose of fire-prone vegetation through cost sharing programs. In fact, the economics of cost-sharing to avoid wildfires compared to the losses caused by those conflagrations compels states and local WUI communities to act.

RECOGNIZE THE MODEL USED FOR CONTRACTING AND OPERATING MECHANIZED EQUIPMENT AND AERIAL FIREFIGHTING ASSETS REQUIRES A COMPLETE TRANSFORMATION

Every organization must access and leverage its core competencies. Strategic missions are only successful when an educated and trained workforce combines with the most advanced physical assets available to accomplish tasks. Federal firefighting agencies lack vital strategies, policies, and practices for mechanized equipment and aerial firefighting assets. The underutilization of these assets increases the likelihood of catastrophic annual wildfire disasters, as demonstrated by the increase in the number of acres destroyed in the last decade. Unless and until wildfire suppression becomes the priority for federal wildfire fighting agencies, we can expect exacerbated wildfire disasters annually.

Mechanized equipment. Heavy equipment to maintain roads off-season appears to be a core competency of federal agencies; however, reliance on bulldozers, excavators, skidgines (a combination of fire engine and a tracked skidder), and feller bunchers to assist in fighting wildfires and managing the forests demands a complete transformation. Our research reveals a lack of use of mechanized equipment and little, if any, training on the equipment in forest conditions. Imagine burning trees and branches falling amid heavy equipment attempting to establish a fire line. Unprotected hand crews working alongside rapidly moving heavy equipment operators require collaboration to coordinate safe and effective forest fire fighting. Absent annual, rigorous, in-the-woods day and nighttime training on heavy equipment, firefighting crews, dispatchers, and incident commanders have little to no knowledge of the best use of this equipment or its capabilities. It's a stretch in credibility to consider heavy equipment use a core competency when the most recent Western states in-the-woods training session was held in 2012. Moreover, the lack of regular coordinated training exercises, including daytime and nighttime scenarios, significantly jeopardizes safety of firefighting crews and heavy equipment teams. Arcane and discriminating contracting practices do not engage the best available resources, such as logging contractors, with the optimal equipment and forest terrain knowledge.

In high-wind conditions, bulldozers and excavators represent the best assets for use during the initial attack. Aerial assets may have limited ability to launch or deploy water or retardant in such conditions; however, regardless of weather conditions or forecasts, we found little staging or prepositioning of bulldozers or excavators by federal or state firefighting agencies. Available wildfire predictive systems provide prepositioning and staging opportunities. Only California, Georgia, and Texas appear more attuned to this practice.

We suggest a much higher priority in the use of mechanized equipment by chartering an organization unit from existing resources. Our assessment suggests federal agency ownership should be the norm for heavy equipment required for road maintenance and other normal forest maintenance activities. For firefighting and forest management missions, federal agencies should contract the services of expert third-party contractors to access the most experienced personnel and advanced, innovative equipment. This practice maximizes financial incentives for efficient operations. Use an extensive investment in predictive systems to stage and preposition contractors and equipment. Remove excessive bureaucratic processes to allow faster response times when the bell rings for an initial attack. Often, an immediate decision to act will, based on history, cost substantially less. We cannot be afraid to act!

Additionally, conduct vendor summits at least every other year for contractors, state and international fire agencies, and forest management personnel. Display the latest technologies and innovations through regular communication between vendors and these parties and provide a forum for an exchange of suggestions for improvement in equipment usage. Currently, no such event exists, demonstrating a lack of agency leadership and disregard of responsibility to assemble the best assets and personnel to manage the nation's forests.

Aerial firefighting equipment. Having a fleet of just 18 air tankers and 108 helicopters to fight wildfires across the U.S. continent is a national embarrassment. If we are serious about fighting wildfires, the fleet must increase substantially. A country like Israel maintains a fleet of 14 single-engine air tankers (SEATs) to cover its 8,550 square miles, yet the United States contracts for only 60 SEATs and 18 air tankers to cover this country's 3.8 *million* square miles.[12] Why such a disparity? Because federal firefighting policy considers fire suppression terrible practice. FS officials constantly espouse that wildfire suppression causes the buildup of more fuel sources for the next wildfire. On the other hand, they refuse to acknowledge reforestation of wildfire impacted forests takes ten to fifteen years to redevelop. In reality, the buildup of fuels is due to the lack of forest management immediately after a wildfire event to clear, restore, and replant areas impacted by wildfires. In our view, the embedded culture of federal firefighting agencies is responsible for the wildfire crisis we face today. Further, the fact that we have 100 million acres of forestland demanding serious attention should cause us to question why these agencies are not held accountable for their total mismanagement.

Despite the substantial investment in satellites, ground sensors, and other technologies that can accurately predict and detect naturally caused wildfire events, we found little to no prepositioning of aerial firefighting assets. As a result, response times for arriving at a wildfire event range from several hours to several days. GAO, RAND Corporation studies, and others indicate that if wildfires are not controlled within the first twenty-four hours, the likelihood exists for larger, costlier, and more complex conflagrations. Response times suffer and the wildfire increases absent an adequate fleet of prepositioned aircraft ready to respond, "when the bell rings."

Our examination of the aerial aircraft fleet and contracting practices leads us to the following conclusions:

- Modify the method of utilizing and contracting for aerial firefighting assets to:

  » Immediately discontinue using expensive, renewable short-term contracts with small independent operators currently providing aerial firefighting which wastes more than $250 million annually.

  » Instead of independent operator contracts, the government should purchase a fleet of 200 SEATs, 75 to 125 large air tankers (LATs), and 30 to 50 very large air tankers (VLATs).

  » Federal and state firefighting agencies should engage private independent operators in long-term contracts to provide pilots, crews, maintenance, and logistics services.

- Develop the ideal next-generation aerial firefighting platform(s) and engage aircraft manufacturers to bid and build aircraft specifically designed for aerial firefighting:

  » The current fleet of LATs and VLATs of converted commercial aircraft is not the most accurate, safest, most cost-efficient, or optimal for fighting wildfires in many WUI and geographic conditions.

  » While the FS touts advanced next-generation technologies, the disparate mixed platforms of the current fleet of converted commercial aircraft fall far short of modern aircraft fleet management practices.

  » Convene a global conference to develop specifications, potential demand, and a time line to achieve aircraft manufacturer cost and delivery efficiencies.

- Fight wildfires twenty-four hours a day, including nighttime, when optimum humidity and temperature conditions exist:

  » Technologies are available to safely fight wildfires during nighttime hours.

  » Training should include mechanized and aerial fighting equipment for the simulated incident.

- Require active competitive bidding before awarding sole-source contracts:

- » Two examples where single suppliers hold long-term sole source agreements are the fire-retardant provider and terrestrial field sensors contractor.

- » Absent multiple sources of supply, potential supply and support risks could impact efforts to suppress or predict wildfire incidents.

- » Neither the breadth of opportunities nor the best cost options should be considered without competitive bidding.

- » Innovation opportunities suffer when markets limit the number of competitors.

- Provide local firefighters and incident command teams with state and federal financial incentives to put fires out within two days. Defined incentives will compensate individuals and teams for their paid overtime compensation and base wage adjustments to calculate retirement pay.

- Require complete transparency and accountability at the federal and state departments for

  - » "let burn" decisions;

  - » state and federal reimbursement of local disaster relief funding, including intergovernmental transfers; and

  - » requests for proposals and final contracts detailing the expected performance standards and specifications for heavy firefighting assets (bulldozers, helicopters, aerial aircraft).

EMBRACE AND ENCOURAGE INNOVATION, RESEARCH, AND DEVELOPMENT

Today's technologies improve predictability of wildfire occurrences and the size and potential outcomes when a wildfire ignites. Ecosystem management appears to be an excellent option for economic forest floor fuels removal, yet technology is not optimized. USDA Forest Service Chief Vicki Christensen assessed the technology outlook in testimony before the House Appropriations Committee on April 15, 2021: "We are also investing in several key technology and modernization portfolios; including, Data Management, Enhanced Real-Time Operating Picture, Decision Support Applications, and Modern Tools for a Modern Response."[13]

Our research detected a number of potentially impactful tools, some of which are under evaluation by wildland fire agencies, to assist in managing wildland fire and fire risk. The following are examples of some intriguing technologies:

1. Use drone imaging and forest inventory assessment technology to enable digitization of every tree in a forest with identifier numbers on each tree. This technology could be a game-changer. Robotic mechanical thinning of the forest, including insect-infested trees, could be accomplished precisely and more economically than current labor-intensive inventory and thinning practices.

2. Rely on military global positioning systems for earlier detection of wildfire outbreaks. In 2018 the FS experimented with the system in California and quickly spotted four episodes.[14]

3. Implement 24-hour, 7-days-per-week, 365-days-per-year drone surveillance over fire-prone areas. These drones contain thermal band sensors, laser, photo, and telecommunications to mobilize firefighting assets quickly.

4. Tap into fire simulation software to identify forest areas with high fuel risk to locate vegetation removal coordinates. Additional simulation software enables forest managers to design precise patterns for planned or prescribed burns to clear surface vegetation without destroying tree canopies.

5. Take advantage of predictive fire perimeter models containing artificial intelligence to anticipate behavior and the anticipated circumference of wildfires. These tools could be valuable in determining where to drop retardant and establish buffer zones in the initial attack stages. Establish a predicted perimeter to control the destruction and remove fuel sources from oncoming flames.

6. Use fire-stopping retardant to extinguish fires. Current retardant serves as a fire break to control the wildfire perimeter but does not extinguish the flames. Recent tests of a new treatment for putting out wildfires are promising but additional refinement is required before federal wildfire agencies can approve it.

7. Use viscoelastic retardant fluids to treat high-risk landscapes. The ecosystem-friendly fluid improves adherence and retention on targeted vegetation and enables prolonged prevention of fire ignition in wildland areas.

8. Employ renovated grinders, loaders, and hauling trucks to reduce the cost to transport woody biomass to environmentally friendly, cogeneration facilities for fuel, heat, and electricity production.

9. Consider available technologies to turn woody biomass into renewable natural gas for producing green heat and electricity.

Unfortunately, our observation of the FS uncovered a culture extremely slow to adopt technology. There is little, if any, capability to communicate timely wildfire behavior current data to on-the-ground firefighting teams. We found little evidence of information sharing systems communicating wildfire behavior through computerized applications in easily understandable formats to firefighting teams on the ground. FedEx and UPS drivers have better handheld remote technologies than our wildfire fighting teams. The bureaucratic apparatus has crawled into the twenty-first century without a coherent, strictly enforced timetable to hire and equip our brave firefighters with the best technologies.

Unique research skill sets exist in our colleges and universities. The FS accesses these resources to conduct many studies, but sometimes findings from duplicate and triplicate studies confuse the initial findings due to insistence on seeking other viewpoints. Recall the bark beetle study conducted in 1930 (see chapter 9) that suggested an eradication plan. More than ninety years later, additional investigations and debates continue without meaningful action other than performing limited forest thinning in higher risk areas. We found no other agency conducting more research studies than the FS. Instead, what if colleges and universities were incentivized to develop innovative technologies? What if the FS allotted a $50,000 grant to a professor/university to develop a cost-effective, environmentally-friendly, tree canopy treatment to avoid supercharged wildfires? What if a genetic or natural means of eradicating insect infestation were discovered and the university/professor could share the resulting patent rights?

It's time for the FS to step it up. Technology provides many potential solutions. Not every research project will bear fruit, but as hockey player Wayne Gretzky stated, "You miss 100 percent of the shots you don't take.[15] It's time to implement technology solutions instead of initiating more studies.

INCLUDE DATA ANALYTICS AND INFORMATION TRANSPARENCY.

The charts and graphs presented in this book represent rudimentary examples of data worthy of ongoing analysis. Astounding volumes of data have been accumulated from decades of statistics collected by the forest and wildfire agencies about wildfires, forest inventories, assets utilized, personnel deployed, and costs incurred. Developing internal expertise to analyze these substantial data repositories is a best practice when managing large organizations. Questions that could be answered include:

- What is the best mix of tree species in a fire-prone forest to defend against wildfires?

- Which aircraft platform has the best cost efficiency and retardant-drop effectiveness?

- Could more carbon be sequestered if a particular tree species were planted and introduced into a recent wildfire-burned area?

- Based on wind conditions for a specific location, what information from previous wildfire events could help model a wildfire perimeter for a current outbreak?

It's time to mine the data, carefully analyze the information, and act on the results. Data analytics represent a powerful and under-utilized opportunity to improve forest management outcomes for these agencies.

Pew Research studies and GAO reports found a lack of transparency and consistent application at the state and federal level when accounting for wildfire costs (see chapter 3). One example: the CRS reports $6.11 billion in 2020 wildfire suppression costs but the total from the National Interagency Fire Center (NIFC) is $2.23 billion (see figure 3 in chapter 3).[16] Which one, if any, is accurate? Another example: Combining "call when needed" with "essential use" contracted assets when reporting aerial firefighting inventories misstates the true number of aircraft available during fire seasons.

Despite 100 million acres of forestland at serious risk due to mismanaged FS maintenance, the latest wildfire crisis strategy projects wildfire risk to be diminished by thinning two million acres per year in the next ten years. Insect infestation as reported by the FS will outpace any proposed thinning effort. Then there is, perhaps, the most blatant of misstatements—the famous public service announcement in which Smokey Bear intones, "Nearly 9 out of every 10 wildfires are caused by humans—only *you* can prevent wildfires."[17] The data actually indicates 40 percent of all wildfire acres in the past twenty years were caused by humans (see figure 1 in chapter 4). What's Smokey been smoking in those woods? Would Smokey and whoever approved this public service announcement please tell us the truth? The reality is that only four out of 10 wildfires are caused by humans.

Professionally run organizations recognize the value of transparency, consistency, and honest reporting of information for developing long-lasting creditability and respect among consumers.

DETERMINE IF AND HOW WILDFIRE SUPPRESSION, FOREST MANAGEMENT, AND CARBON OFFSET REDUCE GREENHOUSE GASES.

When the United States reentered the Paris Accord in early 2021, it agreed to a "net-zero" target for greenhouse gas (GHG) emissions by 2050—a six million metric ton net reduction of GHGs compared to the previous 1990 Kyoto Protocol starting point. On a per capita basis, the United States achieved a 30 percent reduction in GHGs through 2019. Other participants in the Paris Accord also agreed to aggressive targets to reach a "net-zero" GHG position by 2050, meaning any carbon dioxide equivalent (CO_{2e}) emissions into the atmosphere would be sequestered with a corresponding reduction. The current estimated global imbalance is approximately 4.9 gigatons, meaning more CO_{2e} is being emitted into the atmosphere each year than is sequestered.[18]

It is undeniable that this crusade has impacted every country and will continue to do so. To achieve "net-zero targets," forests are critical as they efficiently sequester CO_2 emissions. Fewer global wildfires would significantly impact global GHG emissions while retaining and growing forest carbon stores.

We've noted that wildfires produce about 10 percent of global GHGs every year. Our analysis reveals that in 2020, U.S. wildfires emitted almost 200 million metric tons of CO_2, about five percent of the annual total of emitted United States' GHGs. Suppose wildfires grow due to global warming in the years ahead as many experts believe. The impact of wildfire GHGs would be even worse on the environment, not to mention the carcinogenic fine particulate matter ($PM_{2.5}$) smoke wildfires produce and the public inhales. That alone justifies aggressive wildfire suppression at the forefront of FS policy instead of the minimized approach described in this chapter. Putting fires out produces positive GHG outcomes, but wildfire suppression alone will not reduce the 4.9 gigaton net GHG emissions by 2050 globally.

Reducing deforestation, insect infestations, wildfire destruction, and improved building regulations for urban expansion in forested areas represent the principal drivers for improving forest resilience while promoting higher levels of CO_2 sequestration. There are three potential solutions to improve forest CO_2 sequestration:

- First: provide funding resources that put people and assets to work in the highest risk areas to thin forests, clear forest floor fuels buildup, and remove insect-infested trees.

- Second: enact legislation and regulations strengthening building codes, permits, and buffer zones, particularly in risky WUI areas.

- Third: sell carbon offsets, providing funds to produce desired forest management results through afforestation projects.

The first two options are viable means to achieve these goals. The third, selling carbon offsets, requires careful analysis and perhaps some skepticism. Global concern over a warming planet created an entirely new and unregulated marketplace where carbon offsets and credits are sold, traded, and exchanged. There are few standards on project verification goals or actual carbon sequestration results certification. Land rights and use protections for carbon offsets are not developed fully; neither are conflict resolution practices firmly in place.

Avoided conversion offsets incentivize the project to retain the forest "as is" for up to 100 years. This may not be in the best interest of forest management since it creates a higher, more intense wildfire risk. Cross-border carbon offset and carbon credit transactions call sovereign rights into question. What happens if one country defaults on its carbon agreement after receiving funding? How is this conflict resolved? What if the possible resolution is unsatisfactory to the funding country? How does a country enforce its claim?

Some answers may be forthcoming. In November 2021, as part of the United Nations Climate Framework meetings in Glasgow, a subcommittee released initial guidelines for transacting the exchange of carbon credits and offsets.[19] Meanwhile, the task force charged with overhauling the current voluntarily carbon marketplace is ramping up efforts to improve current blatant shortcomings.

Companies are on the receiving end of growing pressure to develop and achieve a carbon-neutral footprint. This should drive up the demand for projects worldwide, offering opportunities to acquire carbon credits and offsets, particularly for high-quality projects. As supply attempts to meet demand, sovereign forests, such as the United States forestlands, could be put into play by politicians and bureaucrats driven to achieve net-zero goals. In mid-2021, Michigan made a slippery slope policy decision to sell forestry carbon credits, becoming the first state to stop logging on its forest stock (at its Pigeon River Country Forest), presumably allowing forest stock to sequester more carbon.[20] (Currently, federal policy prohibits the sale of any federal forest carbon credit, carbon offset, or similar instrument as a matter of national security interest.) Nevertheless, in late 2021, the FS announced commencement of its Carbon Partnership Program, a worrisome and ominous note. This initiative seeks to explore ways the FS can leverage external demand and partnerships for carbon projects, driven by President Biden's decision to re-enter the Paris Accord.[21]

Imagine this: What if the United States changed policy and offered carbon credits and offsets for our nation's forests? A country seeking to achieve its net-zero goals, China, for example, buys these carbon contracts and commits to the usual 40- or 100-year time frames. Then a catastrophic fire season occurs, destroying the forest under contract. What are the economic, national sovereignty, and potential security threats? And what if, like Michigan, the carbon contracts compensate government entities permitting forests to become dense and at even greater risk of intense and devasting wildfires? All these potential scenarios are in play as global net-zero pressures increase.

Carbon offset and carbon credit markets require significantly more time, deep analysis, and public debate before becoming a significant funding vehicle for forest management practices. Projects must be accretive, not neutral, and measurable, ensuring CO_{2e} sequestration. Economic interests, such as logging, require careful evaluation, mainly when selling carbon credits. Once adequately regulated, robust verification and certification practices should be employed to fund significant projects. Only those contracts assuring protection of our economic, national security, and sovereign rights should be granted. As prominent sixth-century B.C. Chinese philosopher Lao Tzu stated, "Nature does not hurry, yet everything is accomplished."[22]

DEVELOP PRIVATE/PUBLIC PARTNERSHIPS

Private and public partnerships are a viable solution. Our view of America's forest management takes a holistic perspective and suggests a balanced approach. As we have argued throughout this book, extinguish wildfires as quickly as possible with all available resources. Fight each incident as a war effort, engaging the mission's most advanced mechanical and aerial fighting assets. Actively manage forests through scientifically proven practices including thinning, selected prescribed burns, logging, replanting, grazing, and environmentally friendly herbicides. Shortly after a wildfire event, undertake logging to salvage the remaining trees, repair the forest bed, and replant resilient tree species compatible with the existing forest and topography. Conduct forest management efforts regionally, considering the differences in WUI interference, ecosystems, climate, watershed, Indigenous cultures, and wildlife maintenance.

At the same time, we are pragmatic. We understand implementing this strategy requires resources that may be unavailable from already strained federal and state budgets. Costs of restoring 100 million acres of deteriorating forests could range from $40 billion to $80 billion and acquisition of an optimal aerial firefighting fleet is an additional $2 billion to $4 billion. We also recognize the controversies associated with various forest management techniques, such as prescribed burns, grazing, mechanical thinning, environmentally friendly herbicides, lack of financial resources, and the endless debate over which forest management techniques led to the current crisis.

Several recent developments provide a path forward. First, in November 2021, Congress amended the National Environment Protection Act to adopt the Infrastructure Investment and Jobs Act. The amendment allows for reasonable forest management activity up to 3,000 acres (called categorical exclusion) for any project that otherwise would have mandated filing a costly environmental impact study requiring a lengthy review.[23] Ideally, this action will streamline routine forest management and stop the invariable litigation prior to the amendment.

Second, the FS and Bureau of Land Management entered into Good Neighbor Authority (GNA) and Shared Stewardship agreements with nearly every state. In addition, the FS created the Tribal Relations Strategic Plan to increase engagement in shared stewardship of lands on which Indigenous tribes retain legal rights and interests of 4,000 miles of shared boundaries.[24] Goals outlined in these agreements align with regional ecosystem forest management needs using proven practices and advanced scientific research.

Impressive work has already been accomplished. The Rocky Mountain Research Station. partners with tribal nurseries on such projects as restoring native plants to repair the ecosystem, making it more diverse and resilient to wildfires while enhancing plant production for cultural, spiritual, and medicinal purposes.[25] Fire managers established advanced spatial analysis tools to predefine the best fire access and fire control points during the initial attack to combat wildfires effectively. Other projects in the works need incubation periods to prove their long-term, scalable, and beneficial effects.

Third, the current Environmental, Social, and Governance (ESG) movement calls on corporations to examine how they can positively improve their ESG activities. Public companies soon will report an ESG rating filed with federal and state regulators in their filings. Private companies are not exempt since customers, suppliers, financial lending institutions, and the boards of directors will examine a company's ESG rating to ensure compliance with best practices. Congress encourages working with the private sector to help fund Shared Stewardship and GNA initiatives. California's Venado Declaration, issued in November 2021, stated it best: "Partner and collaborate in new and unprecedented ways. Living with fire and making this paradigm shift requires new, unique partnerships and necessitates co-ownership of fire management and shared responsibility across all levels of government and with private, tribal, and community-based groups. The focus should be on shared values, vision, and investments, with a central recognition that fire is a natural and essential part of the California landscape. Redefine and broaden how success is measured."[26]

The FS-initiated Conservation Finance Team encourages pursuit of primarily philanthropic funding for watershed, forest restoration, and conservation projects. The restoration work done to date is impressive through the team's relationship with the Nature Conservancy, a non-governmental organization; however, the size and scale of the financial need for forest management outpace this initiative alone.

The Tax Cuts and Jobs Act of 2017 (Public Law No. 115-97) spurred the creation of investment programs called Opportunity Zones. Influenced in part by this legislation, we recommend establishing Forest Protection Zones (FPZs) to launch the private-public partnerships needed to provide the financial and governance resources now required for year-round forest management.[27] Each FPZ board would be charged with setting policy, managing, and overseeing specific regions to offer year-round holistic, sustainable care of forests and ecosystems to prevent forest loss through maintenance, restoration, and reforestation. These partnerships with federal, state, and tribal agencies would take on individual regional governance and policy responsibilities for federal and state forestland management and wildfire suppression, using existing national and state agency infrastructures and their organizations to implement policy. The FPZs would oversee implementation of Shared Stewardship and GNA actions, enabling comparisons by region of successful and unsuccessful forest management practices, and developing a collection of best practices over time. Each FPZ region could provide the opportunity to incubate forest management practices debated for many of the past decades through FS funded research studies.

Figure 9. Map showing the 10 Forest Health Protection regions, which can be adapted for Forest Protection Zones. (Figure by USDA Forest Service, https://www.fs.fed.us/foresthealth/contact-us/regional-contacts.shtml.)

We believe this regional focus should improve outcomes. Too often, large, centralized organizations lose touch with local needs. Ecosystems vary by region and so do their forest management programs. Nimbler, decentralized, smaller organizations frequently can accomplish more than their larger counterparts. The map below (see fig. 9) presents a possible regional view of initial FPZs.

Under our proposal, each region would follow current FS regional designations. Each FPZ would have a seven-person governing board comprised of an FS representative, a state forestry expert, an Indigenous forestland expert, a forestry expert from a nongovernmental organization (such as the Nature Conservancy, National State Foresters Association, or the New York Declaration on Forests), and three corporate representatives chosen by the companies providing the funding to the FPZ for its projects. Each governing board would select a regional manager who has expertise in forest management and wildfire suppression and is employed by a federal or state agency. Governing FPZ boards would report to the USDA Undersecretary of Natural Resources and Environment (or our proposed successor, the Department of Natural Resources). Other governance issues are also incorporated into the FPZ structure (see footnote).

The initial FPZs would maintain existence for fifteen years, with unalterable sunset provisions established. The term should provide a basis for comparing forest management ecosystems and wildfire suppression practices to determine the most impactful and positive outcomes. If a regional FPZ board acts in ways detrimental to national forestry interests, as determined by the FS (or its proposed successor, the Department of Natural Resources, or DNR), the FS or DNR would have veto power. The FPZ board also could litigate should it disagree with the FS/DNR veto.

Each regional FPZ board will assess and qualify forest improvement projects, establish time lines to implement afforestation and restoration projects, produce analytical support, and approve budgets for each project. The board will monitor progress against the established time lines and budgets. At commencement of each FPZ, an independent assessment will be conducted of current carbon stores, carbon emission, and carbon sequestration metrics. This assessment will be repeated after each FPZ's fifteen-year term.

Every FPZ will determine the need for expanded research and development including creating funds to provide incentives for advanced innovation. Each FPZ would retain its own General Area Coordinating Center (GACC), working with the NIFC to access and make available sufficient resources for regional wildfire incidents.

Tax benefits and carbon credits would be offered to incentivize corporate funding to maximize the resources required, but benefits and credits would be limited until achievement of successful outcomes. We envision two sources of funding:

- Individual corporate contributions. Specifically, each corporate sponsor would receive a tax deduction for each investment along with an investment tax credit of 10 percent for at least $10 million funding and an additional 10 percent tax credit after the project or the fifteen-year FPZ term when a third party verifies the environmental goals achievement. If an FPZ achieves an additional 20 percent carbon storage sequestration (and set it as a goal at the onset) through a combination of reduced wildfires and implementation of better forest management practices, corporate sponsors would receive the additional 10 percent tax credit for all funds invested in that FPZ. A goal for another FPZ might be to eradicate 80 percent of insect infestation, thereby reducing deforestation and forest fragmentation in that region. The plans for each FPZ would be tailored to their individual needs (see fig. 10).

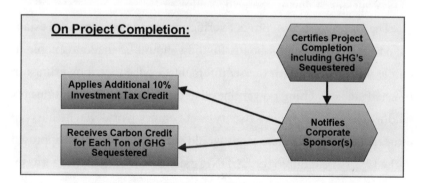

Figure 10. Corporate funding of Forest Protection Zone projects. (Figure created by © Crowbar Research Insights LLC™. All rights reserved.)

Corporate sponsors could access either internal corporate funds or green bonds to invest in FPZ projects approved by the FS or proposed successor DNR. Green bonds focus on ESG initiatives, specifically on the "environmental" or "E" component of ESG. Investments in green bonds can positively influence a company's ESG ratings. Green bonds would be the obligation of corporate sponsors who would invest the entire bond proceeds in specific FPZ projects, such as acquiring mechanized equipment, modern aerial firefighting aircraft to fight wildfires, and performing forest management functions. Unlike other green bonds, we propose that "FS/DNR green bonds" qualify for the deductibility of principal payments when made by the corporate sponsor. In addition, corporate sponsors would receive a 10 percent investment tax credit for any investment greater than $10 million, with immediate application in the year the investment is made. When the FPZ achieves its independently verified environmental improvement goals, the corporate sponsor can claim an additional 10 percent investment tax credit on all funds invested above $10 million on a specific project. If FPZ project goals include reduction of GHGs, one carbon credit will be issued to the corporate sponsor for each metric ton of CO_{2e} sequestered during the project life.

The issuing process, project verification, and tax incentives established by FS/DNR green bonds funding should be made available to public utilities, but with the requirement to bury their powerlines in forested areas. Arching power lines that create sparks and ignitions for wildfires have no place in a twenty-first-century world. Each "burying project" on federal or state lands would require the advanced approval of the FPZ to be eligible for FS/DNR green bonds benefits. Otherwise, the funding necessary to bury powerlines will necessitate rate increases for consumers, which undoubtedly will be prohibitive. Moreover, the tax benefits could defray powerline burying costs for other adjacent private land areas used by the public utility.

In our view, this preferred mechanism incentivizes corporate sponsors and public utilities to earn the carbon credit rather than seek carbon offsets in the murky, unregulated carbon offset and credits marketplace. Moreover, under the FPZ structural design, the governance boards are incentivized to assure that accretive carbon sequestration projects are additive and completed to achieve overall sequestration and environmental goals.

Unfortunately, the federal wildfire agencies and Congress did not include the total funding required in last year's infrastructure bill. We examined our proposed FPZ governance and funding model from various perspectives. Is it perfect? No. It likely will need some modification to become law; however, we could find no other viable incentive-driven alternative to fund the needs of our imperiled forests.

Companies in the S&P 500 held $3.78 trillion in cash and cash equivalents as of September 30, 2021.[28] On the other hand, the U.S. government ran a budget deficit of $2.8 trillion in 2021 and has amassed $31 trillion of the national debt.[29, 30] The funds needed to fix our forest management crisis won't be coming anytime soon from the federal or state governments or the U.S. taxpayers. Investing in our nation's forests could not be a better corporate ESG initiative. Congress, the United Nations, the New York Declaration on Forests, California's Venado Declaration, and the FS encourage creative public-private partnerships. We believe FPZs meet these objectives.[31]

FOCUS ON THE BOTTOM LINE

This book describes an opportunity to make wildfires primarily isolated events rather than annual health, economic, and environmental disasters. We know it is impossible to completely control Mother Nature when a combination of extremely high winds, human behavior, or lightning ignites a forest containing dense underbrush. Instead, we can and must cease the chaos created by our own actions that result in annual loss of life, residences, property, watersheds, and wildlife caused by the mismanagement described throughout the book.

Let's act quickly to enact the FPZ federal and state legislation needed to start this process while consolidating all federal forest management and wildfire suppression functions under one organizational umbrella called the Department of Natural Resources—without lengthy debate or political gamesmanship. Our imperiled forests cannot wait. Americans are tired of government as usual. We have the resources to accomplish this task.

Do we have the will?

CONCLUSION

I went to the woods because I wished to live deliberately, to front only the essential facts of life, and see if I could not learn what it had to teach, and not, when I came to die, discover I had not lived.

—Henry David Thoreau, Walden Pond, 1854[1]

Perhaps the greatest Western naturalist, John Muir, taught us to treasure our forests, soak up their beauty, and refresh our souls from everyday challenges. He championed the establishment of national parks including Yosemite and Sequoia. He understood our lives would be immeasurably crueler unless we care for these national treasures. This famous quote from Muir's book, *Our National Parks*, published in 1901, and the pictures that follow, remind us of the reasons we wrote this book:

> Walk away quietly in any direction and taste the freedom of the mountaineer. Camp out among the grass and gentians of glacial meadows, in craggy garden nooks full of Nature's darlings. Climb the mountains and get their good tidings. Nature's peace will flow into you as sunshine flows into trees. The winds will blow their own freshness into you and the storms their energy, while cares will drop off like autumn leaves. As age comes on, one source of enjoyment after another is closed, but Nature's sources never fail.[2]

The additional John Muir quotes and professional photographs follow all are courtesy of the United States Department of Interior.[3]

"The mountains are calling, and I must go."

Kings Canyon National Park by David Palefsky (www.sharetheexperience.org).

"Between every two pine trees, there is a door leading to a new way of life."

Redwood National and State Parks by Jessica Watz (www.sharetheexperience.org).

CONCLUSION

"I only went out for a walk and finally concluded to stay out till sundown, for going out, I found, was really going in."

Grand Canyon National Park with permission by Robert Shuman (www.sharetheexperience.org).

"Climb the mountains and get their good tidings. Nature's peace will flow into you as sunshine into trees."

Rocky Mountain National Park by Malcolm Boshier (www.sharetheexperience.org).

"Another glorious day, the air as delicious to the lungs as nectar to the tongue."

Shenandoah National Park by N. Lewis, National Park Service. (www.sharetheexperience.org).

"The clearest way into the Universe is through a forest wilderness."

Great Smoky Mountains National Park by Charles Wickham (www.sharetheexperience.org).

CONCLUSION

"How glorious a greeting the sun gives the mountains! To behold this alone is worth the pains of any excursion thousand times over."

Grand Teton National Park by Robert Buman (www.sharetheexperience.org).

"The snow is melting into music."

Denali National Park by Jose Torres (www.sharetheexperience.org).

> "Keep close to Nature's heart . . . and break clear away, once in a while, and climb a mountain or spend a week in the woods. Wash your spirit clean."

Yosemite National Park with permission from Perry Foutch (www.sharetheexperience.org).

> "This grand show is eternal. It is always sunrise somewhere; the dew is never all dried at once; a shower is forever falling; vapor ever rising."

Mount Rainier National Park by Danny Seidman (www.sharetheexperience.org).

APPENDIX

Included in this appendix are selected charts, tables, and graphs developed by the authors to support our conclusions.

CHAPTER 4 — Wildfire Occurrences and Responses

HUMAN-CAUSED WILDFIRES IN FIFTEEN MOST FIRE-PRONE STATES (2010–2021)

	STATE	Area per Fire: GT 300; LT 1,000			Area per Fire: GT 1,000; LT 10,000			Area per Fire: GT 10,000			Total
		Number of Fires	Total Area	Average Area	Number of Fires	Total Area	Average Area	Number of Fires	Total Area	Average Area	Sq. Miles[1]
1	AK	30	17,696	590	19	57,901	3047	10	456,386	45,639	665,384
2	AZ	109	63,361	581	91	293,331	3223	23	1,561,499	67,891	113,990
3	CA	191	98,170	514	160	481,802	3011	66	3,593,280	54,444	163,695
4	CO	72	37,602	522	49	162,170	3310	17	523,552	30,797	104,094
5	FL	208	106,999	514	74	201,260	2720	12	194,982	16,249	65,758
6	GA	33	16,953	514	12	26,308	2192	4	78,654	19,664	59,425
7	ID	132	66,324	502	110	299,506	2723	21	1,163,935	55,425	83,569
8	MT	99	56,031	539	29	290,797	3462	3	227,618	75,873	147,040
9	NV	48	23,808	496	66	218,775	3315	22	1,044,803	47,491	110,572
10	OK	612	323,100	527	269	652,212	2425	16	576,983	36,061	69,899
11	OR	73	36,058	508	71	223,057	3142	24	947,356	39,473	98,379
12	TX	667	317,844	477	367	1,019,841	2779	58	2,353,395	40,576	268,596
13	UT	72	37,898	526	69	194,541	2819	12	406,812	33,901	84,897
14	WA	86	46,620	542	99	310,030	3132	25	1,144,130	45,765	71,298
15	WY	50	26,415	528	43	125,303	2914	11	337,158	30,651	97,813

(Area = Acres; GT = greater than; LT = less than.) ([1] US Census Bureau 2010)

	Totals	2,482	1,274,880		1,528	4,556,834		324	14,610,541		2,204,409
	Avg. per Fire			514			2,982			45,094	

US Total 3,796,742
Percent 58.1%

Source: Human-caused wildfires in the fifteen most fire-prone states. (Table prepared from research performed by Crowbar Research Insights LLC™, using USDA Forest Service wildfire spatial databases, including https://www.fs.usda.gov/rds/archive/Catalog/RDS-2013-0009.5, https://web.archive.org/web/20201202170040/https://gacc.nifc.gov/sacc/predictive/intelligence/NationalLargeIncidentYTDReport.pdf, and https://www.fs.usda.gov/rds/archive/products/RDS-2013-0009.5/_fileindex_RDS-2013-0009.5.html, © Crowbar Research Insights LLC™. All rights reserved.)

LIGHTNING-CAUSED WILDFIRES IN FIFTEEN MOST FIRE-PRONE STATES (2010–2021)

	STATE	Area per Fire: GT 300; LT 1,000			Area per Fire: GT 1,000; LT 10,000			Area per Fire: GT 10,000			Total Sq. Miles[1]
		Number of Fires	Total Area	Average Area	Number of Fires	Total Area	Average Area	Number of Fires	Total Area	Average Area	
1	AK	202	114,170	565	341	1,272,689	3732	252	10,604,281	42,080	665,384
2	AZ	153	86,424	565	249	869,445	3492	60	1,668,357	27,806	113,990
3	CA	74	41,015	554	91	328,570	3611	54	2,545,249	47,134	163,695
4	CO	70	36,005	514	48	79,400	2728	17	485,489	28,558	104,094
5	FL	136	75,708	557	98	357,383	3034	20	617,963	30,898	65,758
6	GA	15	7,131	475	1	4,724	2362	4	366,056	91,514	59,425
7	ID	204	114,376	561	278	943,205	3417	86	4,049,252	47,084	83,569
8	MT	146	83,790	574	177	598,934	3384	54	2,142,392	39,674	147,040
9	NV	103	58,221	565	148	484,877	3276	52	2,067,506	39,760	110,572
10	OK	22	11,575	526	11	29,187	2653	0	-	-	69,899
11	OR	108	62,549	579	163	589,644	3580	78	3,947,926	50,614	98,379
12	TX	198	97,811	494	154	380,982	2474	18	931,823	51,768	268,596
13	UT	83	48,195	581	84	260,509	3139	21	741,512	35,310	84,897
14	WA	50	27,057	541	67	234,420	3552	31	1,439,817	46,446	71,298
15	WY	66	33,167	503	76	244,732	3307	19	617,169	32,483	97,813
	Totals	1,630	897,192		1,986	6,678,701		766	32,224,792		2,204,409
	Avg. per Fire			550			3,363			42,069	

(Area = Acres; GT = greater than; LT = less than.) ([1] US Census Bureau 2010)

US Total 3,796,742
Percent 58.1%

Source: Lightning-caused wildfires in the fifteen most fire-prone states. (Table prepared from research performed by Crowbar Research Insights LLC™, using USDA Forest Service wildfire spatial databases, including https://www.fs.usda.gov/rds/archive/Catalog/RDS-2013-0009.5, https://web.archive.org/web/20201202170040/https://gacc.nifc.gov/sacc/predictive/intelligence/NationalLargeIncidentYTDReport.pdf, and https://www.fs.usda.gov/rds/archive/products/RDS-2013-0009.5/_fileindex_RDS-2013-0009.5.html, © Crowbar Research Insights LLC™. All rights reserved.)

APPENDIX

CHAPTER 5 — Personnel and Mechanized Assets Used to Fight Wildfires

HEAVY EQUIPMENT DESCRIPTION AND ANALYSIS

Small dozers	Effective in building fire line breaks in light fuels on level to moderate terrain. They perform best in soil with few rocks and wet soil conditions, especially when equipped with broad tracks. They can be instrumental in mop-up operations.
Medium-sized dozers	Generally, they are the best overall size for fire line break construction as they are maneuverable and perform well on moderate slopes. They will handle the average fuel and terrain conditions in the mountainous areas and perform well in wet soil conditions when fitted with broad tracks.

All costs listed above are base approximate costs in 2021 without optional equipment and accessories. Current costs likely differ depending on manufacturer and optional accessories included.

Dozers (bulldozers, tractors, CATs) are the most widely recognized mechanized firefighting tool across North America. Cost: $120,000 to $150,000.

Tracked skidders are dozers rigged with chokers/winch or a grapple designed for skidding trees and logs. Cost: $100,000

Pumper CATs are dozers with a water tank, pump, and live hose reel. They function as a tracked skidgine (part dozer and part fire engine) and have fully functioning blades. The chassis is mounted on rigid steel tracks with grousers (cleats) to ensure good traction. Dozers are stable, robust, moderately fast, and versatile. Cost: $130,000 to $160,000 depending on the capacity of the tank.

© Used with permission by Wagner Equipment Co

Feller bunchers are desirable dozers for fire suppression because they can work on slopes up to 60 percent and have a self-leveling cab. They can sit on location and reach a knuckle boom out to the tree or brush. The cutting head usually has clamping or grappling arms that hold on to the material while it is cut, and then the boon repositions the material to place it or bunch it at a rate of about four trees per minute. Cost: $180,000.

Excavators are versatile and maneuverable. They are transported by locations to offload or onload to the lowboy trailers and do not need special locations for offloading or onloading. They offer several functions that can be used for fire suppression including thinning tree canopies and clearing brush while creating a fire line. Cost: $100,000 to $500,000.

(Source of photographs and equipment descriptions: Mechanized Equipment for Fire and Fuels Operations, Valerie Jaffe and Stephen "Obie" O'Brian, 2009, with permission. https://www.wildfirelessons.net/HigherLogic/System/DownloadDocumentFile.ashx?DocumentFileKey=598c5da9-5543-477e-8ec6-017160f3edcb&forceDialog=0.)

Other principal heavy equipment not pictured here includes skidgines (part dozer and part fire engine forwarders (log removal and water pumps to reduce water tender refill times), loaders (tracked machines with 360-degree rotating capability), mulchers and masticators (similar to a wood chipper; it is mounted on an excavator type tractor, which moves through the forest to grind or chip trees and brush, leaving the chips behind or incorporating them into the soil), and road graders. Each can play a significant role in wildfire control and suppression.

Firefighting agencies classify types of dozers basically by size: Type 1 (largest/most potent) through Type 3 (smallest). According to the National Wildfire Coordinating Group's Wildland Fire Incident Management Guide, dozers are limited to a maximum of 75 percent downhill slope and 55 percent uphill. Standard forestry practices state that tracked skidders have an operational ground slope limit between 40 to 50 percent, depending on other site factors. Dozers are used to access ground too steep or rough for wheeled machines or difficult to access sites.

Dozers' most critical terrain hazards are sided slopes over 40 percent, rock, unstable soils, wet areas, and boulders. Dozers with angle and six-way blades cut fire lines, pioneer trails, and push over snags. Most mechanized task forces include a large dozer for quick line pioneering, safety zone construction, and assistance with machine breakdowns. Dozers have faster track speed than tracked excavator-type machines for emergency escape. Safety procedures include two-person crews, portable water, and oxygen sources, radio back-ups, heat level warning systems, flame-resistant glass, air-conditioned, air-filtered cabins.

The table below summarizes the best application of each critical heavy equipment used by forestry personnel today.

Fire Task	Machine Type	Feller Bunchers and Harvesters	Rubber Tired Skidder and Grapple Cable	Dozer and Tracked Skidders
Tree Felling		*	*	*
Brush Cutting		*		
Tree or Log Skidding			*	*
Pruning				
Log Bunching		*	*	*
Log Stacking			*	*
Fire line / Fuelbreak Construction		*	*	*
Water Hauling				*
Water Use				
Emergency Vehicle Recovery			*	*
Site Rehab		*	*	*
Road Work				*
Night Operations		*	*	*

(Table created by Crowbar Research Insights LLC™, with information from Mechanized Equipment for Fire and Fuels Operations, Valerie Jaffe and Stephen "Obie" O'Brian, 2009, with permission. https://www.wildfirelessons.net.)

APPENDIX

Soft Tracks/KMC	Excavators and Tracked Shovel Log	Loaders, Forwarders and Skidgines	Skidgines (Tracked, Rubber-tired)	Mulchers/Masticators	Road Grader/Motor Patrol
	*		*	*	
	*			*	
*	*	*	*		
	*			*	
*	*		*		
*	*	*	*		
*	*	*	*	*	*
		*	*		
			*		
*	*	*	*		*
*	*	*		*	*
	*	*			*
*	*	*	*	*	*

HigherLogic/System/DownloadDocumentFile.ashx?DocumentFileKey=598c-5da9-5543-477e-8ec6-017160f3edcb&forceDialog=0, © Crowbar Research Insights LLC™. All rights reserved.)

CHAPTER 6—Aerial Assets Used to Contain and Extinguish Wildfires

FIXED-WINGED AIR TANKER PROVIDERS AND 2021 EXPECTED PAYMENTS BASED ON ESTIMATED CONTRACT VALUES

We examined the anticipated 2021 contract payments made to the current aerial firefighting community of the aircraft below.

The assumptions are:

- Fuel estimate per gallon is $5.15.

- Aircraft utilization for Essential Use (EU) aircraft is 250 hours above the 160 Mandatory Available Period (MAP) days.

- Aircraft utilization for Call When Needed (CWN) aircraft is 42 hours over 14 days.

- Companies can change their CWN rates at any time due to the arrangements. Companies routinely change their rates.

Fixed Wing 2021 Annual Contract Values Awarded Large Air Tankers and Very Large Air Tankers			
Company	Aircraft	EU/CWN	Hourly rate ($)
10 Tanker	DC-10-30	EU	$41,402
10 Tanker	DC-10-30	EU	$48,193
10 Tanker	DC-10-30	CWN	$41,863
Coulson	C-130-H	EU	$28,169
Coulson	B-737-300	EU	$32,436
Coulson	C-130-H	CWN	$25,414
Aero Air	MD-87	EU	$31,401
Aero Air	MD-87	EU	$31,401
Aero Air	MD-87	EU	$27,224
Aero Air	MD-87	EU	$27,224
Aero Air	MD-87	CWN	$23,503
Aero Flite	RJ-85	EU	$30,468
Aero Flite	RJ-85	EU	$30,468
Aero Flite	RJ-85	EU	$31,474
Aero Flite	RJ-85	EU	$31,474
Aero Flite	RJ-85	EU	$30,659
Aero Flite	RJ-85	EU	$30,659
Aero Flite	RJ-85	CWN	$35,172
Neptune	BAe-146	EU	$30,429
Neptune	BAe-146	EU	$30,429
Neptune	BAe-146	EU	$30,429
Neptune	BAe-146	EU	$30,429
Neptune	BAe-146	CWN	$25,303
Total Values of the Awarded Contracts to Carriers - FY 2021 $149,943,866			

APPENDIX

Contract Cost ($)	Flight Hours	MAP Days	Rev per Company
$10,350,375	250	160	Tanker 10
$12,048,125	250	160	
$1,758,225	42	14	$24,156,725
$7,042,260	250	160	Coulson
$8,108,925	250	160	
$1,067,388	42	14	$16,218,573
$7,850,263	250	160	Aero Air
$7,850,263	250	160	
$6,805,875	250	160	
$6,805,875	250	160	
$987,119	42	14	$30,299,395
$7,616,975	250	160	Aero Flite
$7,616,975	250	160	
$7,868,525	250	160	
$7,868,525	250	160	
$7,664,725	250	160	
$7,664,725	250	160	
$1,477,217	42	14	$47,777,667
$7,607,190	250	160	Neptune
$7,607,190	250	160	
$7,607,190	250	160	
$7,607,190	250	160	
$1,062,746	42	42	$31,491,506

(Table created by Crowbar Research Insights LLC™, Carriers contract award value information to date received from USDA Forest Service, August 24, 2021, © Crowbar Research Insights LLC™. All rights reserved.)

RETARDANT AND OTHER FIRE SUPPRESSION CHEMICAL TOOLS

Other fire-retardant chemicals available for extinguishing or containing wildfires include

- BlazeTamer,

- Eco Fire Solutions' Firewall liquid polymer gel, and

- ICL Performance Products.

Preventive retardants recently received USDA Forest Service (FS) approval. Also developed by Perimeter Solutions, FORTIFY is a ground-based product formulated on a hydrogel platform and polymer technology to provide enhanced durability to weathering and adherence to vegetation. When applied to flammable vegetation and cellulosic material early in the fire season before a wildfire approaches, it provides ongoing protection. It remains effective until a significant rain event of one inch or more.

APPENDIX

PHOS-CHEK FORTIFY being applied from an ATV. (Photograph by USDA Forest Service, public domain.)

Unlike retardant dropped from aircraft, which is colored to help pilots track where the retardant is dropped during active wildfires, this retardant is uncolored and is precision applied from the ground. According to Perimeter Solutions, the FORTIFY product was developed in 2018. During the past three years, utilities, railroads, insurance companies, homeowners, and other industries have adopted it as a solution to prevent wildfire ignitions and proactively protect their property. The FS approved it for use in 2022. It is expected to be introduced into the consumer market in three-quarter gallon containers for home use. The cost per gallon of this new product, however, is approximately $13 per gallon or four times the cost of aerial deployed PHOS-CHEK, currently at $3.10 per gallon, making broad application to forestlands prohibitively expensive. FORTIFY could be spread as a perimeter defense against wildfires for homes in WUI areas.

CHAPTER 7 — Costs and Benefits of Timely Deployment of Assets

SINGLE-ENGINE AIR TANKER (SEAT) COST ANALYSIS

Aircraft Type	Capacity in Gallons	Speed in MPH	Operating Cost per Hour
Air Tractor AT-802A – On wheels	800	180	$3,400
Air Tractor AT-802A – On floats (amphibious)	800	170	$5,300
CL 215-Scooper	1,320	157	Unknown
CL 415-Scooper	1,620	223	Unknown
Grumman S-2T	1,200	235	$4,000

(Table created by Crowbar Research Insights LLC™ with information compiled from: Air Tractor cost estimates are captured from the Air Tractor website. Grumman S-2T cost estimates were derived from Aerialvisuals.ca for the Kansas Grumman S-2T acquired by Ag Air Service, Nickerson, Kansas (2017). Viking Air (CL-215 and 415) data provided by Technical Support, Viking Aircraft, Sidney, BC, Canada, , © Crowbar Research Insights LLC™. All rights reserved.)

HELICOPTER COST BY PLATFORM AND CLASSIFICATION

The following table provides cost estimates for contracted firefighting helicopters. All rates are based on EU hourly rates, although the FS acknowledges that CWN hourly rates are considerably greater than comparable EU rates.

APPENDIX

Helicopter Nomenclature	Capacity in Gallons	Speed in MPH	FS Cost per Hour
Category 1			
Boeing 234 Chinook CH-47	3,000	137	$7,600
Sikorsky S-64 Sky Crane CH-54	2,650	105	$7,900
Eurocopter AS 332L Super Puma	2,000	156	$4,200
Boeing Vertol 107 Sea Knight CH-46	1,100	140	$4,400
Sikorsky S-70 Black Hawk UH-60	1,000	183	$4,250
Sikorsky S-61 Sea King SH-3	1,000	154	$4,500
Kaman K-Max	660	91	$2,200
Category 2			
Bell 412	360	140	$2,300
Bell 212	360	115	$2,200
Bell UH-1 Huey	360 or 324	126	$2,000
Bell 204A	360 or 324	125	$2,000
Category 3			
UH-60 Black Hawk (Military)	360	174	MIL
Alouette 316B	180	115	$2,000
Aerospatiale SA 316B Lama	180	115	$2,000
Bell 407 OH-58D Kiowa Warrior	180	152	$2,000
Eurocopter AS 350 "Squirrel"	180	161	$1,200
Bell 206B Long Ranger	120	115	$1,100
MD-500 OH-6A Cayuse/Loach	120	144	$1,100

(Table created by Crowbar Research Insights LLC™ with military designators compiled by Crowbar Research Insights LLC™. Solicitation No. AG-024B-S-14-9212 U.S. Forest Service Region 2 Call When Needed Type III Helicopter Services National Office, Section "C" various subsections. Military nomenclature compiled by Crowbar Research Insights LLC™ using data from https://gacctest.nifc.gov/rmcc/logistics/aviation/docs/R2Ty3CWN.pdf. NOTE: UH-60 Military helicopters are now commercially available for sale, © Crowbar

Research Insights LLC™. All rights reserved). Military designators compiled by Crowbar Research Insights LLC. All rights reserved.)

The Chinook CH-47s are operated mainly with Bambi Buckets. Recently, three companies—Helimax, Columbia Helicopters, and Billings—have installed retrofitted internal bunkers, allowing these Chinook helicopters to have higher intervention speeds while taking up to 3,200 gallons of water internally. As a result of these modifications, the internal capacity for these Chinooks is greater than most of the fixed-winged large air tankers (LATs).

CURRENT FLEET DATE OF MANUFACTURE AND CURRENT REPLACEMENT AND MARKET VALUES

The following table identifies the manufacturing window of the current fleet and the estimated cost to acquire one aircraft at today's market price.

APPENDIX

Aircraft	Dates Manufactured	2022 Estimated cost of acquisition
McDonnell Douglas DC-10	1971–1981	$2.0
Lockheed C-130-H Model	1954–Current series "J"	N/A
Boeing 737-300 series (WN)	1967–1999	$2.5M
McDonnell Douglas MD-87	1987–1993	$1.0M
Avro RJ-85	1992–2002	$1.7M
British Aerospace BAe-146	1983–2002	$1.5M
Boeing 747-400	1968–1998	$18.4M
DeHavilland Dash-8	1984–1992	Unknown

(Table created by Crowbar Research Insights LLC™, with information compiled from: The AVITAS BlueBook of Jet Aircraft Values first half 2021—with permission of AVITAS, Inc. Chantilly, VA. All rights reserved. Other notes compiled by Crowbar Research Insights LLC, *not covered in AVITAS BlueBook:* Value assumptions: half-life/half-time maintenance and overhaul status, good operating condition, good on all Airworthiness Directives, typical service bulletin incorporation, fulsome records etc. Can be +/–100 percent for variance from these assumptions. Run-out engines alone could eat up 100 percent of these values, © Crowbar Research Insights LLC™. All rights reserved.)

Notes and other sources: The DeHavilland Dash-8 brand was revived by Longview Aviation Capital in 2019. Dates of aircraft manufacture provided by various sources including individual aircraft manufacturers and AVITAS as referenced above (all rights reserved). (WN) designator source identifies Coulson Group aircraft acquisition from Southwest Airlines, see *Puget Sound Business Journal*, May 23, 2017, updated May 26, 2017, 2:50pm PDT (Andrew McIntosh, reporter).

FIXED-WING COST PER MILE—
CONTRACT COMPANIES AND AIRCRAFT

Company	Aircraft	Probability of Success	Capacity (in Gallons)	Success on the Fire
SEAT (water)	AT-802A	.87	800	696
SEAT (retardant)	AT-802A	.87	800	696
10 Tanker	DC-10 EU 2 aircraft	.7	9,400	6,580
	DC-10 CWN 1 aircraft	.7	9,400	6,580
Coulson	C-130 EU 1 aircraft	.68	4,000	2,720
	B-737 EU 1 aircraft	.68	4,000	2,720
	C-130 CWN 1 aircraft	.68	4,000	2,720
Aero Air	MD-87 EU 4 aircraft	.68	3,000	2,040
	MD-87 CWN 1 aircraft	.68	3,000	2,040
Aero Flite (Owned by Conair)	RJ-85 EU 6 aircraft	.68	3,000	2,040
	RJ-85 CWN 1 aircraft	.68	3,000	2,040
Neptune	BAe-146 EU 4 aircraft	.68	3,000	2,040
	BAe-146 CWN 1 aircraft	.68	3,000	2,040
Conair	DHC-8-400 CWN 1 aircraft	.74	2,642	1,955
Logistics Air (Aircraft sold)	B-747	.74	19,200	14,208
Galactica Holdings	B-757	.70	7,000	4,900
Embraer	C-390	.74	4,666	3,453

APPENDIX

(Table created by © Crowbar Research Insights LLC™. All rights reserved).

Number of Sorties to over One Mile	Total Flight Time	Average Hourly Rate	Average Cost per Mile
10	10.0	$3,400	$34,000
10	10.0	$3,400	$58,960
1.2	1.6	$44,798	$71,676
1.2	1.6	$41,863	$66,981
2.9	5.3	$28,169	$149,296
2.9	3.8	$32,436	$123,257
2.9	5.3	$25,414	$134,694
3.9	5.0	$29,313	$146,565
3.9	5.0	$23,503	$117,551
3.9	.43	$30,867	$132,728
3.9	43	$35,172	$151,420
3.9	43	$30,429	$130,845
3.9	43	$25,303	$108,803
4.1	5.7	$25,000	$142,500
.9	1.2	$45,000	$54,000
1.6	1.9	$35,000	$66,500
2.3	2.8	$35,000	$98,000

Assumptions made in the tables above

1. Fire location is 150 miles from the tanker base.
2. Lead aircraft is on-scene prior to the arrival of the tanker aircraft.
3. The tanker aircraft will spend twenty minutes in setup and chemical distribution.
4. The average hourly rate is the average of the aircraft under contract (EU or CWN). The same type of aircraft with the same company have different contract rates.
5. Success on the fire probabilities are furnished by FS.
6. Authors' estimates for Conair, Logistics Air, Galactica Holdings, and Embraer.

For the LATs and VLATs currently available in the market, the B-747 is the most successful and least expensive platform for covering one mile of the fire line based on this analysis. Unfortunately, the FS has made the decision to terminate use of the B-747 for fighting wildfires.

Of the currently available LAT and VLAT fleet, the DC-10 is the most financially viable platform with a carrying capacity of over 9,000 gallons. What is interesting to note is that, according to figures provided by the FS, the 10 Tanker DC-10 on the CWN contract is more economical than the average of the two DC-10s on EU contracts. When examining the MD-87 (Aero Air) and the BAe-146 (Neptune) numbers, results are the exact opposite.

The third most financially viable platform is the B-737 (Coulson). The BAe-146 (Neptune) and the RJ85 (Aero Flite) round out the top current four. The Coulson C-130 and the Aero Air MD-87 fall into last place.

APPENDIX

NEW OR MORE RECENTLY MANUFACTURED AIRCRAFT CAPABLE OF FLYING AIR TANKER MISSIONS

Aircraft Manufacturer	Lift Capacity (in Gallons)	Transit Speed (in mph)	Comments
Embraer KC-390	4,000	350	Currently flown by the Brazilian Air Force. Entering flight testing for Portuguese Air Force.
DeHavilland DHC-515 (Water Scooper)	1,850	Unknown	Expected deliveries to begin 2025. Twenty-five European customers have signed letters of intent to purchase the aircraft pending government-to-government negotiations.
DeHavilland DHC-8-402-AT	2,640	250	DHC-8-402-AT tankers and DHC-8-402MR Multirole aircraft are currently undergoing modification by Conair for use in both roles— tanker and multimission.
Boeing B-757	Unknown - TBD	350	Under market evaluation for aerial firefighting use
Airbus 320/330 Family	Unknown - TBD	350	Under market evaluation for aerial firefighting use
Stavatti SM-100AT	Unknown - TBD	Unknown	Under market evaluation Earliest entry likely in 2026

(Table created by © Crowbar Research Insights LLC™. All rights reserved).

It is critical to recognize that current LAT and VLAT platforms may be incapable of navigating to establish fire lines tucked into valleys or up steep slopes. Smaller and more agile airborne platforms are needed. At the same time, water- and retardant-dropping capacity provides the most impact in establishing a defensible fire line. The table above summarizes other air tanker platforms recently released or under evaluation.

The De Havilland DHC-8/400 [DHC 8/400T and DHC 8/400 MR] air tanker and multirole were available during the 2021 fire season in limited numbers. Air Spray (owned by Conair) is modifying eight aircraft for firefighting missions. Versatile, functional, and available in the market, the DHC-8 platform could become a viable air tanker option although its water/retardant drop capacity is somewhat limiting.

The above analysis highlights the lack of a global initiative to develop a genuinely next-generation aerial firefighter. The worldwide need is apparent, and demand would allow economies of scale for the manufacturer ultimately selected.

We view this as an enormous opportunity to develop the specifications based on each phase, from an initial attack to a fully engaged wildfire. FS leadership could play a pivotal role in engaging the global firefighting agency community to develop a next-generation aircraft platform solely as an aerial firefighter.

ENDNOTES

CHAPTER ONE

1. "The Rising Cost of Wildfire Operations: Effects on the Forest Service's Non-Fire Work," United States Department of Agriculture, last modified August 4, 2014. https://www.fs.usda.gov/sites/default/files/2015-Rising-Cost-Wildfire-Operations.pdf.

2. "NASA Moderate Resolution Imaging Spectroradiometer (MODIS), Analysis-Global Fire Data," Global Fire Emissions Database, accessed January 6, 2022, http://www.globalfiredata.org/analysis.html.

3. "UN/ DESA Policy Brief #111: Wildfires – A Growing Concern for Sustainable Development," United Nations Department of Economic and Social Affairs, last modified September 1, 2021, https://www.un.org/development/desa/dpad/publication/un-desa-policy-brief-111-wildfires-a-growing-concern-for-sustainable-development/.

4. "Wildfire Statistics," Congressional Research Service, last modified May 2, 2022, https://fas.org/sgp/crs/misc/IF10244.pdf.

5. "Greetings from the 2020 Wildfire Season," NFPA Journal, last modified November 1, 2020, https://www.nfpa.org/News-and-Research/Publications-and-media/NFPA-Journal/2020/November-December-2020/Features/Wildfire.

6. Andrew C. Scott, *Burning Planet* (Oxford: Oxford University Press, 2018).

7. Jenn R. Marlon, et al, "Climate and human influences on global biomass burning over the past two millennia," *Nature Geoscience* 1, no. 10 (2008), accessed January 14, 2020, https://web.archive.org/web/20080927051047/http:/pmr.uoregon.edu/science-and-innovation/uo-research-news/research-news-2008/september-2008/climate-change-human-activity-and-wildfires-1/.

8. "Wildfire Statistics."

9. "2020 Wildfire Season."

ENDNOTES

10. "Facts + Statistics: Wildfires," Insurance Information Institute, accessed January 17, 2021, https://www.iii.org/fact-statistic/facts-statistics-wildfires.

11. "3 of the largest wildfires in Colorado history have occurred in 2020." NBC News, last modified October 23, 2020, https://www.nbcnews.com/news/us-news/3-largest-wildfires-colorado-history-have-occurred-2020-n1244525.

12. "How 2020 Has Affected the Way We Should Manage Forest Fires," 5280 Magazine, last modified December 15, 2020, www.5280.com/2020/12/how-2020-has-affected-the-way-we-should-manage-forest-fires/.

13. "Top 20 Largest California Wildfires," CAL FIRE, accessed January 17, 2021, https://www.fire.ca.gov/media/4jandlhh/top20_acres.pdf.

14. "2020 Incident Archive," CAL FIRE, accessed January 17, 2021, https://www.fire.ca.gov/incidents/2020/.

15. NIFC Incident Reports and State Forestry websites for California, Oregon, Washington, Idaho, and Montana, Crowbar Research Insights LLC™ Analysis.

16. "Climate Changes in the United States," NASA earth observatory, accessed January 14, 2021, https://earthobservatory.nasa.gov/images/83624/climate-changes-in-the-united-states#:~:text=Since%20consistent%20record%2Dkeeping%20began,the%20United%20States%20was%202012.

17. "A tiny pest helped stoke this year's devasting wildfires," National Geographic, last modified October 1, 2020, https://www.nationalgeographic.com/science/2020/10/bark-beetles-helped-stoke-2020-devastating-wildfires/.

18. Jessica E. Halofsky, David L. Peterson, and Brian J. Harvey, "Changing wildfire, changing forests: the effects of climate change on fire regimes and vegetation in the Pacific Northwest, USA," *Fire Ecology* 16, No. 4 (2020), https://fireecology.springeropen.com/articles/10.1186/s42408-019-0062-8.

19. "Reforestation After the Fire," Oregon State University, accessed May 7, 2022, https://extension.oregonstate.edu/forests/cutting-selling/reforestation-after-fire.

20. "Five Burning Questions about Tree Planting and Wildfire," National Forest Foundation, accessed May 7, 2022, https://www.nationalforests.org/blog/five-burning-questions-about-tree-planting-and-wildfire.

21. "The State of the World's Forests 2020," Food and Agriculture Organization of the United Nations, accessed January 17, 2021, http://www.fao.org/state-of-forests/en/.

22. "Wildfire Statistics."

23. "Wildfire Statistics."

24. "The State of the World's Forests."

25. C. Volker, et al, "Rapid growth of the US wildland-urban interface raises wildfire risk," *PNAS* 115, no. 13 (2018), https://www.pnas.org/content/115/13/3314.short.

26. "Wildland Urban Interface (WUI)," U.S. Fire Administration, accessed January 17, 2021, https://www.usfa.fema.gov/wui/.

27. "Alerts & Notices," United Stated Department of Agriculture Forest Service, accessed January 14, 2021, https://www.fs.usda.gov/wps/portal/fsinternet/cs/list/!ut/p/z0/04_Sj9CPykssy0xPLMnMz0vMAfIjo8zijQwgwNHCwN_DI8z-P0BcqYKBfkO2oCAAgGWNx/?ss=110103&navtype=BROWSEBYSUBJECT&cid=FSEPRD517997&navid=120000000000000&pnavid=null&position=Notices*&ttype=list&counter=61&pname=Forest%20Service%20-%20Alerts.

28. "Where There's Fire, There's Smoke: Wildfire Smoke Affects Communities Distant from Deadly Flames," NRDC Issue Brief, last modified October 2013, https://www.nrdc.org/sites/default/files/wildfire-smoke-IB.pdf.

29. "Ambient (outdoor) air pollution," World Health Organization, last modified May 2, 2018, https://www.who.int/news-room/fact-sheets/detail/ambient-(outdoor)-air-quality-and-health.

30. "Copernicus tracks effects of Arctic Circle wildfires," Copernicus, last modified July 16, 2019, https://atmosphere.copernicus.eu/copernicus-tracks-effects-arctic-circle-wildfires.

31. "CAMS monitors unprecedented wildfires in the Artic," Copernicus, last modified July 11, 2019, https://atmosphere.copernicus.eu/cams-monitors-unprecedented-wildfires-arctic.

32. Gary A. Morris et al, "Alaskan and Canadian forest fires exacerbate ozone pollution over Houston, Texas, on 19 and 20 July 2004," *Journal of Geophysical Research* 111, No. D24 (2006), https://agupubs.onlinelibrary.wiley.com/doi/full/10.1029/2006JD007090.

33. Seung Hyun Lucia Woo et al, "Air pollution from wildfires and human health vulnerability in Alaskan communities under climate change,"

Environmental Research Letters 15, No. 9 (2020), https://iopscience.iop.org/article/10.1088/1748-9326/ab9270.

34. "Smoke from forest fires kills approximately 340,000 people each year," Science for Environment Policy, last modified July 26, 2012, https://ec.europa.eu/environment/integration/research/newsalert/pdf/294na2rss_en.pdf.

35. Fay H. Johnston, et al, "Estimated global mortality attributable to smoke from landscape fires," *Environmental Health Perspective* 120, no. 5 (2012), https://pubmed.ncbi.nlm.nih.gov/22456494/.

36. William Fletcher and Craig Smith, *Reaching Net Zero: What It Takes to Solve the Global Climate Crisis* (Amsterdam: Elsevier, 2020).

37. Guido R. van der Werf et al, "Global fire emissions estimates during 1997-2016," *Earth System Science Data* 9 (2017), https://essd.copernicus.org/articles/9/697/2017/.

38. "Copernicus: 2021 saw widespread wildfire devastation and new regional emission records broken," Copernicus, last modified December 6, 2021, https://atmosphere.copernicus.eu/copernicus-2021-saw-widespread-wildfire-devastation-and-new-regional-emission-records-broken.

CHAPTER TWO

1. "Ronald Reagan—*A Time for Choosing (aka "The Speech")*," American Rhetoric, last modified August 16, 2018, https://www.americanrhetoric.com/speeches/ronaldreaganatimeforchoosing.htm.

2. "Bicentennial Edition: Historical Statistics of the United States from colonial times to 1970, Part 1," United States Census Bureau, accessed June 24, 2021, https://www.census.gov/library/publications/1975/compendia/hist_stats_colonial-1970.html.

3. Michael Williams, *Americans and their forests: historical geography.* (New York: Cambridge University Press, 1989).

4. "Re-inventing the United States Forest Service: Evolution from Custodial Management, to Production Forestry, to Ecosystem Management," Food and Agriculture Organization of the United Nations, Regional Office for Asia and the Pacific, accessed June 24, 2021, http://www.fao.org/3/ai412e/AI412E06.htm.

5. "U.S. Forest Service Home Page," United States Forest Service, accessed January 21, 2021, https://www.fs.usda.gov/.

6. "U.S. Forest Service – About the Agency," United States Forest Service, last modified August 21, 2020, https://www.fs.usda.gov/about-agency.

7. "USDA Forest Service FY 2022 Budget Justification," United States Department of Agriculture, last modified December 2021, https://www.fs.usda.gov/sites/default/files/usfs-fy-2022-budget-justification.pdf.

8. "USDA FS FY 2022 Budget Justification."

9. "Downsizing the Federal Government—Reforming the Forest Service," Cato Institute, last modified July 12, 2016, https://www.downsizinggovernment.org/agriculture/forest-service#:~:text=The%20%20Forest%20Transfer%20Act,150%20million%20acres%20of%20land.

10. "Go Ahead: Move the Forest Service to Interior," Forest Service Employees for Environmental Ethics, last modified October 6, 2017, https://www.fseee.org/2017/10/06/go-ahead-move-the-forest-service-to-interior/.

11. "Federal Land Management: Observations on a Possible Move of the Forest Service into the Department of the Interior," US Government Accountability Office, last modified February 11, 2009, https://www.gao.gov/products/GAO-09-223.

12. "Wildland Fire: Plans and Policy—Fire Management Planning," National Park Service, last modified April 9, 2020, https://www.nps.gov/subjects/fire/wildland-fire-plans-and-policy.htm.

13. "Fire—Wildland Fire Strategic Plan 2020-2024," National Park Service, last modified April 9, 2020, https://www.nps.gov/subjects/fire/wildland-fire-strategic-plan.htm.

14. "The United States Department of the Interior Budget Justifications and Performance Information Fiscal Year 2022—National Park Service," United States Department of the Interior, accessed January 10, 2022, https://www.doi.gov/sites/doi.gov/files/fy2022-nps-budget-justification.pdf.

15. "US Department of the Interior Indian Affairs—Branch of Wildland Fire Management," US Department of the Interior Indian Affairs, accessed January 21, 2021, https://www.bia.gov/bia/ots/dfwfm/bwfm.

ENDNOTES

16. "US Department of the Interior Indian Affairs—Division of Forestry and Wildland Fire Management," US Department of the Interior Indian Affairs, accessed January 21, 2021, https://www.bia.gov/bia/ots/dfwfm.

17. "US Department of the Interior Bureau of Land Management—Fire and Aviation Program," US Department of the Interior Bureau of Land Management, accessed January 21, 2021, https://www.blm.gov/programs/fire-and-aviation.

18. "US Department of the Interior Bureau of Land Management—FY 2022 Interior Budget in Brief," US Department of the Interior Bureau of Land Management, accessed January 10, 2022, https://www.blm.gov/about/budget#:~:text=The%202022%20budget%20proposes%20program%20increases%20totaling%20%24160,of%20the%20Nation's%20land%20and%20water%20by%202030.

19. "US Fish & Wildlife Service—Fire Management," US Fish & Wildlife Service, accessed January 10, 2022, https://www.fws.gov/program/fire-management.

20. "The United States Department of the Interior Budget Justifications and Performance Information Fiscal Year 2022—Fish and Wildlife Service," United States Department of the Interior, accessed January 10, 2022, https://www.doi.gov/sites/doi.gov/files/fy2022-fws-budget-justification.pdf.

21. "The United States Department of the Interior Budget Justifications."

22. "US Department of the Interior—Office of Wildland Fire Home Page," US Department of the Interior, accessed January 21, 2021, https://www.doi.gov/wildlandfire.

23. "US Department of the Interior—Office of Wildland Fire—Budget." US Department of the Interior accessed January 21, 2021, https://www.doi.gov/wildlandfire/budget.

24. "US Department of the Interior Bureau of Land Management—Fire and Aviation Program," US Department of the Interior Bureau of Land Management, accessed January 21, 2021, https://www.blm.gov/programs/fire-and-aviation.

25. "US Department of the Interior Bureau of Land Management—FY 2022 Interior Budget in Brief," US Department of the Interior Bureau of Land Management, accessed January 10, 2022, https://www.doi.gov/budget/appropriations/2022/highlights.

26. "US Fire Administration Home Page," US Fire Administration, last modified January 20, 2021, https://www.usfa.fema.gov/.

ENDNOTES

27. "FY 2022 Budget in Brief—Homeland Security," Homeland Security, accessed January 10, 2022, https://www.dhs.gov/sites/default/files/publications/dhs_bib_-_web_version_-_final_508.pdf.

28. "A Guide to the Disaster Declaration Process and Federal Disaster Assistance," FEMA, accessed February 4, 2021, https://www.fema.gov/pdf/rrr/dec_proc.pdf.

29. "How a Disaster Gets Declared," FEMA, last modified November 25, 2020, https://www.fema.gov/disasters/how-declared.

30. "National Interagency Fire Center—What is NIFC?," National Interagency Fire Center, accessed January 21, 2021, https://www.nifc.gov/about-us/what-is-nifc.

31. "National Interagency Coordination Center Home Page," National Interagency Coordination Center, accessed January 21, 2021, https://www.nifc.gov/nicc/index.htm.

32. "National Multi-Agency Coordinating Group Operations Plan 2020," National Multi-Agency Coordinating Group, accessed January 21, 2021, https://www.nifc.gov/nicc/administrative/nmac/NMACOpsPlan.pdf.

33. "Forests and Rangelands—Wildland Fire Leadership Council," Forest and Rangelands, accessed January 21, 2021, https://www.forestsandrangelands.gov/leadership/index.shtml.

34. "Interagency Standards for Fire and Fire Aviation Operations 2020," Department of the Interior, accessed January 21, 2021, https://www.nifc.gov/standards/guides/red-book.

35. Wildland Fire Leadership Council—National Strategic Committee Charter, Forests and Rangelands, accessed January 21, 2021, https://www.forestsandrangelands.gov/documents/leadership/wflc/meetings/NSC_Charter20150303.pdf.

36. "National Wildfire Coordinating Group—Mission," National Wildfire Coordinating Group, accessed January 21, 2021, https://www.nwcg.gov/.

37. "Master Interagency Agreement For Wildland Fire Management," US Department of the Interior and US Department of Agriculture, last modified February 10, 2017, https://www.nwcg.gov/sites/default/files/docs/eb-master-ia-for-wildland-fire-mgmt.pdf.

38. "Where's There's Fire, There's Smoke: Wildfire Smoke Affects Communities Distant from Deadly Flames," National Resources Defense Council, accessed June 24, 2021, https://www.nrdc.org/sites/default/files/wildfire-smoke-IB.pdf.

39. "Federal Land Ownership: Overview and Data," Congressional Research Service, last modified February 21, 2020, https://fas.org/sgp/crs/misc/R42346.pdf.

40. "Public Land Ownership in the United States," Headwaters Economics, last modified June 17, 2019, https://headwaterseconomics.org/public-lands/protected-lands/public-land-ownership-in-the-us/.

41. Chelsea Pennick McIver, Philip S. Cook, and Dennis R. Becker, "The Fiscal Burden of Wildfires: State Expenditures and Funding Mechanisms for Wildfire Suppression in the Western U.S. and Implications for Federal Policy," *State and Local Government Review*, December 6, 2021, https://journals.sagepub.com/doi/full/10.1177/0160323X211061353.

42. "National Association of State Foresters Home Page," National Association of State Foresters, accessed January 21, 2021, https://www.stateforesters.org/.

43. "Gardening Trends for 2019," University of Vermont Department of Plant and Soil Science, last accessed June 24, 2021, http://pss.uvm.edu/ppp/articles/trends19.html.

44. "Welcome to the HABITAT NETWORK website," Habitat Network, last accessed June 24, 2021, https://www.habitatnetwork.org/. [Crowbar Research Insights LLC™ calculation $503 annual spend divided by 0.225 acres, equals $2,236 yearly spend per acre.]

45. "Milton Friedman on the minimum wage," American Experiment, last modified May 4, 2018, https://www.americanexperiment.org/milton-friedman-on-the-minimum-wage/.

CHAPTER THREE

1. "1 billion Acres at Risk for Catastrophic Wildfires, U.S. Forest Service Warns," NPR, last modified June 5, 2019, https://www.npr.org/2019/06/05/729720938/1-billion-acres-at-risk-for-catastrophic-wildfires-us-forest-service-warns.

2. "Federal Wildfire Management: Ten-Year Funding Trends and Issues (FY2011-FY2020)," Congressional Research Service, last modified October 28, 2020, https://crsreports.congress.gov/product/pdf/R/R46583.

3. "Suppression Costs-Federal Firefighting Costs (Suppression Only)," National Interagency Fire Center, accessed June 29, 2021, https://www.nifc.gov/fire-information/statistics/suppression-costs.

4. D.V. Spracklen, et al., "Impacts of climate change from 2000 to 2050 on wildfire activity and carbonaceous aerosol concentrations in the western United States," *Journal of Geophysical Research* 114, (2009), accessed June 29, 2021, https://agupubs.onlinelibrary.wiley.com/doi/epdf/10.1029/2008JD010966.

5. Yongqiang Liu, et al., "Future US wildfire potential trends projected using a dynamically downscaled climate change scenario," *Forest Ecology and Management* 294, (2013), accessed June 29, 2021, https://www.sciencedirect.com/science/article/abs/pii/S037811271200388X.

6. "Facts + Statistics: Wildfires," Insurance Information Institute, accessed June 29, 2021, https://www.iii.org/fact-statistic/facts-statistics-wildfires.

7. "Large-loss fires in the United States," National Fire Protection Association, accessed June 29, 2021, https://www.nfpa.org/News-and-Research/Data-research-and-tools/US-Fire-Problem/Large-loss-fires-in-the-United-States.

8. "Facts + Statistics: Wildfires."

9. "High Cost of Wildfire Insurance Hurts California Home Sales," The Wall Street Journal, last modified January 25, 2020, https://www.wsj.com/articles/high-cost-of-wildfire-insurance-hurts-california-home-sales-11578220200.

10. "Record hurricane season and major wildfire – The natural disaster figures for 2020," Munich Re, last modified January 7, 2021, https://www.munichre.com/en/company/media-relations/media-information-and-corporate-news/media-information/2021/2020-natural-disasters-balance.html#-848711503.

11. "The natural disasters of 2018 in figures," Munich Re, last modified August 1, 2019, https://www.munichre.com/topics-online/en/climate-change-and-natural-disasters/natural-disasters/the-natural-disasters-of-2018-in-figures.html.

12. "Natural catastrophes in 2020," Swiss Re Institute, last modified March 30, 2021, https://www.swissre.com/institute/research/sigma-research/sigma-2021-01.html.

13. "California utilities to spend billions to cut wildfire risk," KRON4, last modified February 6, 2021, https://www.kron4.com/news/california/california-utilities-to-spend-billions-to-cut-wildfire-risk/.

14. "PG&E bills to rise amid electricity upgrades to combat wildfires," The Mercury News, last modified January 1, 2021, https://www.mercurynews.com/2020/12/31/pge-bills-rise-january-2021-electricity-system-upgrades-gas-wildfire/.

15. "Xcel Energy seeking a Colorado rate hike to fund grid wildfire maintenance," Complete Colorado – Page Two, last modified August 8, 2020, https://pagetwo.completecolorado.com/2020/08/08/xcel-energy-seeking-a-colorado-rate-hike-to-fund-grid-wildfire-maintenance/.

16. "Xcel Energy requests $17.2 million funded by rate increases for Colorado wildfire mitigation efforts," The Center Square, last modified October 27, 2020, https://www.thecentersquare.com/colorado/xcel-energy-requests-17-2-million-funded-by-rate-increases-for-colorado-wildfire-mitigation-efforts/article_df8de5f2-147c-11eb-bfdf-4f193987d4c7.html.

17. "PG&E plan to bury power lines gets price tag; who will foot the bill?" The Mercury News, last updated February 11, 2022, https://www.mercurynews.com/2022/02/10/pge-plan-to-bury-power-lines-gets-more-expensive-who-will-foot-the-bill/#:~:text=The%20company%20announced%20to%20stockholders,to%20%243.75%20million%20per%20mile.

18. "The Full Community Costs of Wildfire," Headwaters Economics, accessed June 29, 2021, https://headwaterseconomics.org/wp-content/uploads/full-wildfire-costs-report.pdf.

19. "What We Don't Know About State Spending on Natural Disasters Could Cost Us," The Pew Charitable Trusts, last modified June 19, 2018, https://www.pewtrusts.org/en/research-and-analysis/reports/2018/06/19/what-we-dont-know-about-state-spending-on-natural-disasters-could-cost-us.

20. "How States Pay for Natural Disasters in an Era of Rising Costs," Pew Charitable Trusts, last modified May 12, 2020, https://www.pewtrusts.org/en/research-and-analysis/reports/2020/05/how-states-pay-for-natural-disasters-in-an-era-of-rising-costs.

21. "Budgeting for Disasters-Approaches to Budgeting for Disasters in Selected States," United States Government Accountability Office, last accessed June 29, 2021, https://www.gao.gov/assets/670/669362.pdf.

ENDNOTES

22. "Disaster Relief Fund: Monthly Report as of December 31, 2020 [also 2016-2019]," Federal Emergency Management Agency, last modified January 8, 2021, https://www.fema.gov/sites/default/files/documents/fema_jan-2021-disaster-relief-fund-report.pdf.

23. Chelsea Pennick McIver, Philip S. Cook, and Dennis R. Becker, "The Fiscal Burden of Wildfires: State Expenditures and Funding Mechanisms for Wildfire Suppression in the Western U.S. and Implications for Federal Policy," *State and Local Government Review*, December 6, 2021, https://journals.sagepub.com/doi/full/10.1177/0160323X211061353.

24. "US Fire Department Profile 2019," National Fire Protection Association, last modified December 2021, https://www.nfpa.org/-/media/Files/News-and-Research/Fire-statistics-and-reports/Emergency-responders/osfdprofile.pdf.

25. "Economy-Wide Statistics Division," United States Census Bureau, last modified April 14, 2015, https://www2.census.gov/about/partners/sdc/events/steering-committee/2015-04/2015-medina.pdf.

26. "Overall Wildfire Activity Reported to the NICC," National Interagency Fire Center, last accessed January 9, 2022, https://www.predictiveservices.nifc.gov/intelligence/2020_statssumm/wildfire_charts_tables20.pdf.

27. "Orange County Fire Authority – FY 2019/20 Adopted Budget," Orange County Fire Authority, accessed June 29, 2021, https://ocfa.org/Uploads/Transparency/OCFA%202019-2020%20Adopted%20Budget.pdf.

28. "The Full Community Costs."

29. "The Full Community Costs."

30. "The Full Community Costs."

31. "Billion-Dollar Weather and Climate Disaster: Summary Stats," NOAA-National Centers for Environmental Information, accessed June 29, 2021, https://www.ncdc.noaa.gov/billions/summary-stats.

32. "The Costs and Losses of Wildfires," National Institute of Standards and Technology, last modified November 2, 2017, https://www.nist.gov/publications/costs-and-losses-wildfires.

33. "California Wildfires Are Driving Up Fire Insurance Premiums—and Home Prices," Mansion Global, last modified October 13, 2020, https://www.mansionglobal.com/

articles/california-wildfires-are-driving-up-fire-insurance-premiums-and-home-prices-220244.

34. "CO2 Emissions from Commercial Aviation," The International Council on Clean Transportation, last modified October 2020, https://theicct.org/sites/default/files/publications/CO2-commercial-aviation-oct2020.pdf.

35. "How Wildfires Can Affect Climate Change (and Vice Versa)," Inside Climate News, last modified August 23, 2018, https://insideclimatenews.org/news/23082018/extreme-wildfires-climate-change-global-warming-air-pollution-fire-management-black-carbon-co2/.

CHAPTER FOUR

1. "Wildland Fire Management – Improvements Needed in Information, Collaboration, and Planning to Enhance Federal Fire Aviation Program Success," United States Government Accountability Office, last modified August 2013, https://www.gao.gov/assets/gao-13-684.pdf.

2. "Wildfire Statistics," Congressional Research Service, last modified May 2, 2022, https://fas.org/sgp/crs/misc/IF10244.pdf.

3. "Fighting Wildfires," Public Broadcasting Service, accessed February 25, 2021, https://www.pbs.org/wgbh/americanexperience/features/burn-fighting-wildfires/.

4. "Wildfire and Acres," National Interagency Fire Center, accessed July 2, 2021, https://www.nifc.gov/fireInfo/fireInfo_stats_totalFires.html.

5. "The story of Smokey Bear," US Department of Agriculture Forest Service, last modified August 4, 2014, https://www.fs.usda.gov/features/story-smokey-bear.

6. "Wildfire Statistics."

7. "Wildfire Statistics."

8. "Statistics," National Interagency Fire Center, accessed July 2, 2021, https://www.nifc.gov/fireInfo/fireInfo_statistics.html.

9. "Colorado's Marshall Fire expected to be 10th-costliest in US history," FOX31 Denver KDVR-TV, last modified January 3, 2022, https://kdvr.com/news/data/colorados-marshall-fire-expected-to-be-tenth-costliest-in-u-s-history/.

ENDNOTES

10. "Handcrews," US Department of Agriculture Forest Service, accessed February 25, 2021, https://www.fs.usda.gov/science-technology/fire/people/handcrews.

11. "Hotshots," US Department of Agriculture Forest Service, accessed February 25, 2021, https://www.fs.usda.gov/science-technology/fire/people/hotshots.

12. "Engine Crews," US Department of Agriculture Forest Service, accessed February 25, 2021, https://www.fs.usda.gov/science-technology/fire/engine-crews.

13. "Smokejumpers," US Department of Agriculture Forest Service, accessed February 25, 2021, https://www.fs.usda.gov/science-technology/fire/people/smokejumpers.

14. "Helitack," US Department of Agriculture Forest Service, accessed February 25, 2021, https://www.fs.usda.gov/science-technology/fire/helitack.

15. "Equipment and Tools," US Department of Agriculture Forest Service, accessed February 25, 2021, https://www.fs.usda.gov/science-technology/fire/equipment-tools.

16. "Responding to Wildfires," Forest Service—National Interagency Fire Center, last modified December 27, 2017, https://www.youtube.com/watch?v=maT-G67YdaKY&ab_channel=ForestService-NationalInteragencyFireCenter.

17. "Branch of Wildland Fire Management," US Department of the Interior Indian Affairs, accessed February 25, 2021, https://www.bia.gov/bia/ots/dfwfm/bwfm.

18. "Aviation," US Department of the Interior Indian Affairs, accessed February 25, 2021, https://www.bia.gov/bia/ots/dfwfm/bwfm/responding-wildfires/aviation.

19. "Helitack Program," US Department of the Interior Indian Affairs, accessed February 25, 2021, https://www.bia.gov/bia/ots/dfwfm/bwfm/helitack-program.

20. "Wildland Fire Engine Program," US Department of the Interior Indian Affairs, accessed February 25, 2021, https://www.bia.gov/bia/ots/dfwfm/bwfm/responding-wildfires/wildfire-engine-program.

21. "Hand Crews," US Department of the Interior Indian Affairs, accessed February 25, 2021, https://www.bia.gov/bia/ots/dfwfm/bwfm/responding-wildfires/hotshot.

ENDNOTES

22. "BLM Fire Program," US Department of the Interior Bureau of Land Management, accessed February 25, 2021, https://www.blm.gov/programs/fire.

23. "What We Do," US Department of the Interior Bureau of Land Management, accessed February 25, 2021, https://www.blm.gov/programs/public-safety-and-fire/fire-and-aviation/about-fire-and-aviation/what-we-do.

24. "National Interagency Coordination Center (NICC)," National Interagency Fire Center, accessed July 2, 2021, https://www.nifc.gov/about-us/what-is-nifc/nicc.

25. "National Incident Management System," FEMA, accessed February 25, 2021, https://www.fema.gov/emergency-managers/nims.

26. "National Response Framework," FEMA, accessed February 25, 2021, https://www.fema.gov/emergency-managers/national-preparedness/frameworks/response.

27. "2021 National Interagency Mobilization Guide – Chapter 10," National Interagency Coordination Center, accessed February 25, 2021, https://www.nifc.gov/nicc/mobguide/Chapter%2010.pdf.

28. "Course Introduction: IS-700.b An Introduction to the National Incident Management System," FEMA, accessed February 25, 2021, https://emilms.fema.gov/is_0700b/curriculum/1.html.

29. "Wildland Fire: Incident Command System Levels," National Park Service, last modified February 13, 2017, https://www.nps.gov/articles/wildland-fire-incident-command-system-levels.htm.

30. "IS-011.c: An Introduction to the Incident Command System, ICS 100," FEMA, accessed February 25, 2021, https://emilms.fema.gov/is_0100c/groups/133.html.

31. "Wildland Fire: Incident Command System," National Park Service, last modified February 13, 2017, https://www.nps.gov/articles/wildland-fire-incident-command-system.htm.

32. "National Incident Management System," FEMA, accessed February 25, 2021, https://www.fema.gov/sites/default/files/2020-07/fema_nims_doctrine-2017.pdf.

ENDNOTES

33. "Initial Attack Dispatcher," National Wildfire Coordinating Group, last modified January 18, 2022, https://www.nwcg.gov/positions/iadp/position-qualification-requirements.

34. "Wildland Fire Management – Improvements Needed in Information, Collaboration, and Planning to Enhance Federal Fire Aviation Program Success," United States Government Accountability Office, last modified August 2013, https://www.gao.gov/assets/gao-13-684.pdf.

35. "National Fire Weather Annual Operating Plan 2021," National Weather Service, accessed July 2, 2021, https://www.weather.gov/media/fire/2021_Natl_AOP.pdf.

36. "National Significant Wildland Fire Potential Outlook," National Interagency Coordination Center, accessed February 25, 2021, https://www.nifc.gov/nicc/predictive/outlooks/outlooks.htm.

37. "Leadership In Decision Support Services – Predictive Services Program Overview," National Interagency Coordination Center, accessed February 25, 2021, https://www.predictiveservices.nifc.gov/.

38. "Predictive Services," US Department of Agriculture Forest Service, accessed February 25, 2021, https://www.fs.fed.us/r6/fire/success/predictive-services/.

39. "Fire Forecasting," US Department of Agriculture Forest Service, accessed February 25, 2021, https://www.fs.usda.gov/science-technology/fire/forecasting.

40. "Statement of Victoria Christiansen, Chief of the USDA Forest Service, Before the House Committee on Appropriations," US House of Representatives Document Repository, last modified April 15, 2021, https://docs.house.gov/meetings/AP/AP06/20210415/111425/HHRG-117-AP06-Wstate-ChristiansenV-20210415.pdf.

41. "NWCG Airtanker Base Directory," National Wildfire Coordinating Group, last modified January 31, 2022, https://www.nwcg.gov/publications/507.

42. "Firefighting Aircraft Recognition Guide," CAL FIRE, accessed February 5, 2022, https://www.fire.ca.gov/media/8592/aviationguide_final_web-booklet.pdf.

CHAPTER FIVE

1. "Greetings from the 2020 Wildfire Season," NFPA Journal, accessed July 2, 2021, https://www.nfpa.org/News-and-Research/Publications-and-media/NFPA-Journal/2020/November-December-2020/Features/Wildfire.

2. "Hiring and retention in the US Forest Services is a growing issues," Wildfire Today, last modified May 20, 2021, https://wildfiretoday.com/2021/05/20/hiring-and-retention-in-the-us-forest-service-is-a-growing-issue/.

3. "The History of McCulloch," McCulloch, last accessed May 15, 2021, https://www.mcculloch.com/int/discover/history/.

4. "The Modern Bulldozer—A Forest Service Project," National Park Service History, last modified October 15, 2010, http://npshistory.com/publications/usfs/region/1/early-days/4/sec28.htm.

5. "USDA FY 2022 Budget Justification," United States Department of Agriculture, last modified December 2021, https://www.fs.usda.gov/sites/default/files/usfs-fy-2022-budget-justification.pdf.

CHAPTER SIX

1. "Great Aviation Quotes – Air Power," Aviation Quotations, accessed June 17, 2022, https://www.aviationquotations.com/airpowerquotes.html.

2. "Firefighting air tankers, the early years," Fire Aviation, last modified January 22, 2021, https://fireaviation.com/2021/01/22/firefighting-air-tankers-the-early-years/.

3. Information received by Crowbar Research Insights LLC™ from John Yount, former pilot with the USDA Forest Service, on August 22, 2020.

4. "Confronting The Wildfire Crisis," USDA Forest Service, last modified January 2022, https://www.fs.usda.gov/sites/default/files/Confronting-Wildfire-Crisis.pdf.

5. "Firefighting aircraft 'increasingly ineffective' amid worsening wildfires," FireRescue1, last modified April 7, 2019, https://www.firerescue1.com/flame-retardants/articles/firefighting-aircraft-increasingly-ineffective-amid-worsening-wildfires-qSG7F3n5Y8SyQIVc/.

ENDNOTES

6. "Statement of Victoria Christiansen, Chief of the USDA Forest Service, Before the House Committee on Appropriations," US House of Representatives Document Repository, last modified April 15, 2021, https://docs.house.gov/meetings/AP/AP06/20210415/111425/HHRG-117-AP06-Wstate-ChristiansenV-20210415.pdf.

7. "Manitoba converts to contractors for operating the province's air tankers," Fire Aviation, last modified November 29, 2018, https://fireaviation.com/2018/11/29/manitoba-converts-to-contractors-for-operating-the-provinces-air-tankers/.

8. "Contracting for Fire," Northern Rockies Coordinating Center, accessed June 17, 2022, https://gacc.nifc.gov/nrcc/dispatch/equipment_supplies/agree-contract/steps.htm.

9. "Cost of exclusive use vs. call when needed air tankers," Fire Aviation, last modified February 21, 2018, https://fireaviation.com/2018/02/21/cost-exclusive-use-vs-call-needed-air-tankers/.

10. "Cost of exclusive use."

11. "Cost of exclusive use."

12. "Big water-bombing aircraft en route to Australia to fight fires delayed by international disasters," ABC News, last modified January 15, 2020, https://www.abc.net.au/news/2020-01-15/firefighting-aircraft-delayed-by-international-disasters/11869676.

13. "CAL FIRE Home Page," CAL FIRE, accessed June 17, 2022, https://www.fire.ca.gov/.

14. "NWCG Airtanker Base Directory," National Wildfire Coordinating Group, last modified January 31, 2022, https://www.nwcg.gov/publications/507.

15. "Aerial Firefighting Use and Effectiveness (AFUE) Report," USDA, last modified March 2020, https://www.fs.usda.gov/sites/default/files/2020-08/08242020_afue_final_report.pdf.

16. "USDA Forest Service FY 2022 Budget Justification," United States Department of Agriculture, last modified December 2021, https://www.fs.usda.gov/sites/default/files/usfs-fy-2022-budget-justification.pdf.

17. "Take the Fight to the Night," 60 Minutes-CBS News, last modified September 26, 2021, https://www.cbsnews.com/video/60minutes-2021-09-26/.

ENDNOTES

18. "Forest Service says there's not enough time to issue new helicopter contracts for 2022," Fire Aviation, last modified December 17, 2021, https://fireaviation.com/2021/12/17/forest-service-says-theres-not-enough-time-to-issue-new-helicopter-contracts-for-2022/.

19. "Air Tractor Aerial Fire-Fighting Solutions—Fire Agency Briefing," Valley Air Briefing, last modified September 2015, https://www.dnr.wa.gov/publications/rp_fire_aviation_fireboss_agencybriefing.pdf.

20. "Air Tractor Aerial Fire-Fighting Solutions."

21. Email to Crowbar Research Insights LLC™ principals from the Bureau of Land Management personnel at National Interagency Fire Center, dated April 22, 2022.

22. "World's largest firefighting plane grounded as the West braces for another destructive wildfire season," CNN, last modified April 28, 2021, https://www.cnn.com/2021/04/28/weather/global-supertanker-wildfire-grounded-trnd-scn-wx/index.html.

23. "World's largest firefighting plane grounded."

24. "Forest Service to abandon Coast Guard HC-130H program," Fire Aviation, last modified February 16, 2018, https://fireaviation.com/2018/02/16/forest-service-abandon-coast-guard-hc-130h-program/.

25. "Firefighting Aircraft Recognition Guide," CAL FIRE, accessed June 17, 2022, https://gacc.nifc.gov/swcc/dc/azpdc/operations/documents/aircraft/links/Aircraft%20Recognition%20Guide.pdf.

26. Geniy V. Kuznetsov, et al., "Rates of high-temperature evaporation of promising fire-extinguishing liquid droplets," *Applied Sciences* 9, no 23 (2019), accessed June 17, 2022, https://www.mdpi.com/2076-3417/9/23/5190/htm.

27. "Standards for Airtanker Operations," USDA, last modified July 2019, https://www.fs.usda.gov/sites/default/files/2019-09/standards_for_airtanker_operations_-_final_-_2019_approved_0.pdf.

28. "Wildland Fire Incident Management Field Guide," National Wildfire Coordinating Group, last modified January 2014, http://npshistory.com/publications/fire/pms-210.pdf.

29. "Chain (unit)," Wikipedia, last modified September 27, 2021, https://simple.wikipedia.org/wiki/Chain_(unit).

30. "Air Attack Against Wildfires—Understanding U.S. Forest Service Requirements for Large Aircraft," The RAND Corporation, last modified 2021, https://www.rand.org/content/dam/rand/pubs/monographs/2012/RAND_MG1234.pdf.

CHAPTER SEVEN

1. "Dalai Lama Quotes," BrainyQuote, accessed May 18, 2022, https://www.brainyquote.com/quotes/dalai_lama_386414.

2. The United States Department of the Interior Budget Justifications and Performance Information Fiscal Year 2021—Wildland Fire Management," United States Department of the Interior, accessed May 18, 2022, https://www.doi.gov/sites/doi.gov/files/uploads/fy2021-budget-justification-wfm.pdf.

3. "Air Attack Against Wildfires—Understanding U.S. Forest Service Requirements for Large Aircraft," RAND Corporation, Homeland Security and Defense Center, last modified 2012, https://www.rand.org/content/dam/rand/pubs/monographs/2012/RAND_MG1234.pdf.

4. See Appendix, Chapter 4, data used from Natural Caused Chart, combining the annual average number of wildfire incidents greater than 300 and less than 10,000 acres and wildfire incidents greater than 10,000 acres.

5. "National Interagency Coordination Center—Wildland Fire Summary and Statistics Annual Report 2020," National Interagency Coordination Center, accessed May 18 2022, https://www.predictiveservices.nifc.gov/intelligence/2020_statssumm/annual_report_2020.pdf.

6. "Forest Service currently has several advertisements posted for fire aviation services," Fire Aviation, last modified May 23, 2019, https://fireaviation.com/2019/05/23/forest-service-currently-has-several-advertisements-posted-for-fire-aviation-services/.

7. Information received by Crowbar Research Insights LLC™ from Bureau of Land Management on April 22, 2022.

8. "CAL FIRE Home Page," CAL FIRE, accessed May 18, 2022, www.fire.CA.gov.

9. "CAL FIRE Home Page."

ENDNOTES

10. "DynCorp International Lands $352M Contract to Help CAL FIRE," DynCorp International, last modified November 9, 2020, https://potomacofficersclub.com/dyncorp-international-lands-352m-contract-to-help-cal-fire/.

11. "Israel adds to their air tanker fleet," Fire Aviation, last modified January 8, 2015, https://fireaviation.com/2015/01/08/israel-adds-to-their-air-tanker-fleet/.

12. "Air Attack Against Wildfires"

13. "NWCG Airtanker Base Directory," National Wildfire Coordinating Group, accessed May 18, 2022, https://www.nwcg.gov/publications/507.

14. "Aviation Annual Report," U.S. Forest Service, last modified January 1, 2021, https://www.fs.usda.gov/sites/default/files/2021-06/CY2020_USFSAviationReport_Final_1.pdf.

15. "Encyclopedia Britannica," Britannica, accessed May 18, 2022, https://www.britannica.com/.

CHAPTER EIGHT

1. "Reviewing the horrid global 2020 wildfire season," Yale Climate Connections, last modified January 4, 2021, https://yaleclimateconnections.org/2021/01/reviewing-the-horrid-global-2020-wildfire-season/.

2. "Fires, Forests and the Future: A Crisis Raging Out of Control?" World Wide Fund for Nature, accessed July 14, 2021, https://wwf.panda.org/discover/our_focus/forests_practice/forest_publications_news_and_reports/fires_forests/.

3. "24+ Blazing Wildfire Statistics for the US and Abroad," Policy Advice, last modified February 5, 2021, https://policyadvice.net/insurance/insights/wildfire-statistics/.

4. "Forests Ablaze. Causes and effects of global forest fires," World Wildlife Fund, accessed March 13, 2021, https://www.wwf.de/fileadmin/fm-wwf/Publikationen-PDF/WWF-Study-Forests-Ablaze.pdf.

5. "Area burned in 2019 forest fires in Indonesia exceeds 2018 – official," Reuters, last modified October 21, 2019, https://www.reuters.com/article/us-southeast-asia-haze-area-burned-in-2019-forest-fires-in-indonesia-exceeds-2018-official-idUSKBN1X00VU.

ENDNOTES

6. "Indonesia fires cost nation $5 billion this year: World Bank," Mongabay, last modified December 20, 2019, https://news.mongabay.com/2019/12/indonesia-fires-cost-nation-5-billion-this-year-world-bank/.

7. "Forests Ablaze"

8. "KEPCO may face damage suit for forest fire in Gangwon Province," The Korean Times, last modified May 9, 2019, https://www.koreatimes.co.kr/www/tech/2019/04/694_266763.html.

9. "2020 Northern Thailand forest fires snapshot," World Wildlife Fund, last modified April 15, 2020, https://www.wwf.or.th/en/?uNewsID=362337.

10. "Forest Fires Rage on Southeast Asian Peninsula," Asia Sentinel, last modified April 30, 2020, https://www.asiasentinel.com/p/forest-fires-rage-on-southeast-asian.

11. "Fire Resistant Tropical Forest on Brink of Disappearance," Swansea University, last modified December 18, 2020, https://www.swansea.ac.uk/press-office/news-events/news/2020/12/fire-resistant-tropical-forest-at-brink-of-disappearance-in-indonesian-regions-due-to-human-modification.php.

12. Raban Dutta, et al. "Big data integration shows Australian bushfire frequency is increasing significantly," *Royal Society Open Science* 3, (2016), https://royalsocietypublishing.org/doi/full/10.1098/rsos.150241.

13. "Six Months After Australia's Wildfires, Recovery Continues," Direct Relief, last modified June 24, 2020, https://www.directrelief.org/2020/06/six-months-after-australias-wildfires-recovery-continues/#:~:text=The%20 2019%E2%80%932020%20Australian%20bushfire,killing%20at%20least%20 34%20people.

14. "24+ Blazing Wildfire Statistics."

15. "3 billion animals killed or displaced in Black Summer bushfires, study estimates," ABC NEWS, last modified July 2, 2020, https://www.abc.net.au/news/2020-07-28/3-billion-animals-killed-displaced-in-fires-wwf-study/12497976.

16. "Smoke from Australian bushfires was more deadly than the fires themselves," Air Quality News, last modified March 25, 2020, https://airqualitynews.com/2020/03/25/smoke-from-australian-bushfires-was-more-deadly-than-the-fires-themselves/#:~:text=Due%20to%20the%20fires%2C%20parts,may%20 experience%20serious%20health%20effects.

ENDNOTES

17. "Australia bushfire smoke travels 12,000 km to Chile," Dateline, last modified July 1, 2020, https://www.sbs.com.au/news/dateline/australia-bushfire-smoke-travels-12-000-kms-to-chile.

18. "Bushfires Release Over Half Australia's Annual Carbon Emissions," Time, last modified December 23, 2019, https://time.com/5754990/australia-carbon-emissions-fires/.

19. "Yes, Australia has always had bushfires: but 2019 is like nothing we've seen before," The Guardian, last modified December 24, 2019, https://www.theguardian.com/australia-news/2019/dec/25/factcheck-why-australias-monster-2019-bushfires-are-unprecedented.

20. "Global 2020 wildfire season"

21. "About us," NSW Rural Fire Service, accessed March 13, 2021, https://www.rfs.nsw.gov.au/about-us.

22. "Fleet," National Aerial Firefighting Centre, accessed March 13, 2021, https://www.nafc.org.au/?page_id=168.

23. "Australia fires: What's being done to fight the flames?" BBC News, last modified January 23, 2020, https://www.bbc.com/news/world-australia-51008051#:~:text=Fire%20crews%20are%20using%20a,it%20has%20more%20than%2060.

24. "NSW Buys Boing 737 Large Air Tanker for Firefighting," Australian Aviation, last modified May 16, 2019, https://australianaviation.com.au/2019/05/nsw-buys-boeing-737-large-air-tanker-for-firefighting/.

25. "National Aerial Firefighting Strategy 2021-26," Australasia Fire and Emergency Service Authorities Council and National Aerial Firefighting Centre, last modified June 2021, https://www.nafc.org.au/wp-content/uploads/2021/07/NAFF_Strategy_Webversion_2021-07-30_v1.1.pdf.

26. "Wildfire Management in Europe: Final Report and Recommendation Paper," CMINE Task Group Wildfire, last modified February 2020, https://www.driver-project.eu/wp-content/uploads/2020/02/DRIVER-CMINE-Wildfire-Report-FINAL-120220.pdf.

27. "Forests Ablaze" https://www.wwf.de/fileadmin/fm-wwf/Publikationen-PDF/WWF-Study-Forests-Ablaze.pdf

28. "24+ Blazing Wildfire Statistics."

ENDNOTES

29. "Why is Portugal so prone to wildfires?" Phys.Org, last modified July 24, 2019, https://phys.org/news/2019-07-portugal-prone-wildfires.html.

30. "Spain battles biggest wildfires in 20 years as heatwave grips Europe," The Guardian, last modified June 27, 2019, https://www.theguardian.com/world/2019/jun/27/hundreds-of-firefighters-tackle-blaze-in-north-east-spain.

31. "Wildfire Management in Europe"

32. "Aerial firefighting resources in Europe," AirMed&Rescue, last modified April 13, 2021, https://www.airmedandrescue.com/latest/long-read/aerial-firefighting-resources-europe.

33. "Portugal to Acquire Five Embraer KC-390s," Aviation Today, last modified July 17, 2019, https://www.aviationtoday.com/2019/07/17/kc-390-portugal/.

34. "Diverse knowledge informing fire policy and biodiversity conservation," Global Paleofire Working Group 2, last modified December 2019, http://pastglobalchanges.org/download/docs/working_groups/paleofire/gpwg2-fire-policy-brief-2019.pdf.

35. "Germany Makes a National Commitment to Rescue Its Forests," Bloomberg, last modified September 27, 2019, https://www.bloomberg.com/news/articles/2019-09-27/germany-s-800-million-plan-to-save-its-forests.

36. "Forests Ablaze"

37. "Federal Agency for Forestry," The Russian Government, accessed March 25, 2021, http://government.ru/en/department/245/.

38. "24+ Blazing Wildfire Statistics."

39. "Forests Ablaze"

40. "Global 2020 wildfire season"

41. "Wildfires in Siberia have burned down an area larger than Greece," CBS News, last modified July 21, 2020, https://www.cbsnews.com/news/wildfires-sibera-russia-burned-area-larger-than-greece-heat-wave/.

42. "Around the world, a fire crisis flares up, fueled by human actions," Mongabay, last modified September 4, 2020, https://news.mongabay.com/2020/09/around-the-world-a-fire-crisis-flares-up-fueled-by-human-actions/

ENDNOTES

43. "Record breaking fires in Siberia," Greenpeace International, last modified August 18, 2021, https://www.greenpeace.org/international/story/49171/russia-record-breaking-fires-siberia/.

44. "'Total destruction': why fires are tearing across South America," The Guardian, last modified October 9, 2020, https://www.theguardian.com/environment/2020/oct/09/a-continent-ablaze-why-fires-are-tearing-across-south-america.

45. "Amazon fires: Are they worse this year than before?" BBC News, last modified August 29, 2020, https://www.bbc.com/news/world-latin-america-53893161.

46. "Wildfires are blazing through the Amazon rainforest in record numbers, burning through a tropical forest vital to countering climate change," Business Insider, last modified August 21, 2019, https://www.businessinsider.com/amazon-rainforest-experiencing-record-number-of-wildfires-this-year-2019-8.

47. "Brazilian Amazon fires scientifically linked to 2019 deforestation: report," Mongabay, last modified September 11, 2019, https://news.mongabay.com/2019/09/brazilian-amazon-fires-scientifically-linked-to-2019-deforestation-report/.

48. "Amazon fires"

49. "Total destruction"

50. "Brazil Deploys Troops to Fight Wildfires After Widespread Outrage at Bolsonaro's Government," Gizmodo, last modified August 25, 2019, https://earther.gizmodo.com/brazil-deploys-troops-to-fight-wildfires-after-widespre-1837560002.

51. "Diverse knowledge informing fire policy"

52. "Brazil's New Forest Code puts vast areas of protected Amazon Forest at Risk, Mongabay, last modified March 4, 2019, https://news.mongabay.com/2019/03/brazils-new-forest-code-puts-vast-areas-of-protected-amazon-forest-at-risk/.

53. "Brazil's Climate Overture to Biden: Pay Us Not to Raze Amazon," The Wall Street Journal, last modified April 21, 2021, https://www.wsj.com/articles/brazils-climate-overture-to-biden-pay-us-not-to-raze-amazon-11618997400.

54. "FAO Publishes Key Findings of Global Forest Resources Assessments 2020," Forest2Market, last modified May 19, 2020, https://www.forest2market.

com/blog/fao-publishes-key-findings-of-global-forest-resources-assessments-2020#:~:text=Key%20Findings%20in%202020&text=The%20world%20has%20a%20total,the%20world%27s%20peoples%20or%20geographically.

55. "Africa is the 'fire continent,' but blazes differ from Amazon," Associated Press, last modified August 28, 2019, https://apnews.com/article/49f74a56b-5564cae8d49f3022223d003.

56. "Explainer: How climate change is affecting wildfires around the world," CarbonBrief, last modified July 14, 2020, https://www.carbonbrief.org/explainer-how-climate-change-is-affecting-wildfires-around-the-world.

57. "Amazon versus Africa Forest fires: Is the world really ablaze?" Deutsche Welle, last modified August 30, 2019, https://www.dw.com/en/amazon-versus-africa-forest-fires-is-the-world-really-ablaze/a-50229553.

58. "South Africa wildfire that burned University of Cape Town, the library of African antiquities is under control," The Washington Post, last modified April 19, 2021, https://www.washingtonpost.com/world/2021/04/18/south-africa-fire-university-cape-town/.

59. Butz RJ. "Traditional fire management: historical fire regimes and land use change in pastoral East Africa." *International Journal of Wildland Fire* 18, no. 4 (2009): 442-450, https://www.researchgate.net/publication/248885433_Traditional_fire_management_historical_fire_regimes_and_land_use_change_in_pastoral_East_Africa.

60. "The Staggering Value of Forests—and How to Save Them," Boston Consulting Group, last modified June 9, 2020, https://www.bcg.com/publications/2020/the-staggering-value-of-forests-and-how-to-save-them.

61. "Effective Wildfire Management: Pathway for sustainable livelihoods and biodiversity conservation in Africa," Society for Conservation Biology, accessed April 15, 2021, https://conbio.org/policy/wildfire-management-sustainable-livelihoods-and-biodiversity-conservation-i.

62. "Wildfires: Information & Facts," Canadian Red Cross, last modified March 7, 2021, https://www.redcross.ca/how-we-help/emergencies-and-disasters-in-canada/types-of-emergencies/wildfires/wildfires-information-facts.

63. "Forests Ablaze"

64. "24+ Blazing Wildfire Statistics."

ENDNOTES

65. "A brief look at some of Canada's biggest wildfires in the last two decades," National Post, last modified January 7, 2020, https://nationalpost.com/pmn/news-pmn/canada-news-pmn/a-brief-look-at-some-of-canadas-biggest-wildfires-in-the-last-two-decades.

66. "Canada's biggest wildfires"

67. Pinno, Bradley, et. al. "Young jack pine and high severity fire combine to create potentially expansive areas of understocked forest." *Forest Ecology and Management* 310, (2013): 517-522, https://www.sciencedirect.com/science/article/abs/pii/S0378112713005975.

68. "Cost of wildland fire protection," Government of Canada, last modified July 15, 2020, https://www.nrcan.gc.ca/climate-change/impacts-adaptations/climate-change-impacts-forests/forest-change-indicators/cost-fire-protection/17783.

69. "Forests Ablaze"

70. "Global Forest Resources Assessments"

71. "Global Fires," Future Earth, accessed March 13, 2021, https://futureearth.org/publications/issue-briefs-2/global-fires/.

72. "Balancing forests and development: Addressing infrastructure and extractive industries, promoting sustainable livelihoods," Forest Declaration Platform, last modified November 2020, https://forestdeclaration.org/wp-content/uploads/2021/10/2020NYDFReport.pdf.

73. Grantham, H.S., et. al. "Anthropogenic modification of forests means only 40% of remaining forests have high ecosystem integrity," *Nature Communications* 11, (2020), https://www.nature.com/articles/s41467-020-19493-3.

74. "Fires, Forests and the Future"

75. "Diverse knowledge informing fire policy"

76. "Global risk of wildfires on the rise as the climate warms, study says," Carbon Brief, last modified July 14, 2015, https://www.carbonbrief.org/global-risk-of-wildfires-on-the-rise-as-the-climate-warms-study-says.

77. "Global Fires"

78. "Forests Ablaze"

79. "Wildfire Management in Europe"

80. "German Forests—Forests for Nature and People," German Federal Ministry of Food and Agriculture, last modified April 2021, https://www.bmel.de/SharedDocs/Downloads/EN/Publications/german-forests.pdf?__blob=publicationFile&v=7.

81. Table 3 Sources: https://www.fao.org/forest-resources-assessment/en/; https://en.wikipedia.org/wiki/2021_Canadian_federal_budget; https://www.statista.com/statistics/455540/china-expenditure-on-agriculture-forestry-and-water-resource-projects/#statisticContainer; https://www.fao.org/3/x5364e/x5364e05.htm; https://www.statista.com/statistics/275337/government-revenue-and-spending-in-portugal/; https://gfmc.online/2021/04-2021/andalucia-and-portugal-collaborate-in-project-to-fight-forest-fires.html; https://en.wikipedia.org/wiki/Economy_of_Portugal#Government_expenditure_by_function; https://theowp.org/greek-pm-approves-500-million-euro-budget-for-wildfire-relief-reforestation/; https://news.mongabay.com/2017/02/indonesian-government-moves-farther-from-community-forestry-target/#:~:text=In%202015%2C%20the%20ministry%20had,165%20billion%20rupiah%20in%202017.; https://www.climatechangenews.com/2021/09/20/indonesia-ends-forest-protection-deal-norway-raising-deforestation-fears/; https://www.ran.org/indonesian-rainforests/; https://meduza.io/en/news/2020/11/26/russian-state-duma-adopts-federal-budget-for-2021-2023; https://www.usnews.com/news/world/articles/2021-11-09/russia-comes-in-from-cold-on-climate-launches-forest-plan; https://efi.int/sites/default/files/files/publication-bank/2020/efi_wsctu_11_2020.pdf; https://www.reuters.com/article/brazil-economy-spending/update-2-brazils-government-cuts-2021-budget-deficit-forecast-idUSL1N2OY1ZX; https://www.reuters.com/article/us-brazil-environment/brazil-proposes-cuts-to-2021-budget-for-environmental-protection-as-deforestation-spikes-idUSKBN29U26S; https://www.latinnews.com/component/k2/item/86267-in-brief-venezuela-approves-increased-state-budget-for-2021.html; https://news.mongabay.com/2019/10/venezuelan-crisis-government-censors-environmental-and-scientific-data/; https://en.wikipedia.org/wiki/List_of_countries_by_government_budget; https://www.exchangerates.org.uk/historical/USD/30_06_2021.

CHAPTER NINE

1. Von Carlowitz, Hans Carl. *Sylvicultura Oeconomica*. Leipzig: Johann Friedrich Bruans, 1713.

ENDNOTES

2. "The State of the World's Forests 2020," Food and Agriculture Organization of the United Nations, accessed August 18, 2021, http://www.fao.org/state-of-forests/en/.

3. "Fast Facts: Canada's Boreal Forest," The Pew Charitable Trusts, last modified March 19, 2015, https://www.pewtrusts.org/en/research-and-analysis/articles/2015/03/19/fast-facts-canadas-boreal-forest.

4. "State Of the World's Forests"

5. "State Of the World's Forests"

6. John A. Stanturf, et al., "Plantation Forests in the United States of America: Past, Present, and Future," *XII World Forestry Congress* (2003), http://www.fao.org/3/xii/0325-b1.htm.

7. "Timber Harvesting on Federal Lands," Congressional Research Service, last modified July 28, 2021, https://crsreports.congress.gov/product/pdf/R/R45688.

8. "Forest Resources of the United States, 1997," US Department of Agriculture, Forest Service, last modified August 2001, https://www.nrs.fs.fed.us/pubs/gtr/gtr_nc219.pdf.

9. "Forest Resources"

10. "U.S. Forest Resource Facts and Historical Trends," United States Department of Agriculture, accessed August 19, 2021, https://www.fia.fs.fed.us/library/brochures/docs/2012/ForestFacts_1952-2012_English.pdf.

11. "Major Trends, Forest Inventory & Analysis, 2002," U.S. Forest Service, accessed August 19, 2021, https://www.fia.fs.fed.us/slides/major-trends.pdf.

12. "Forest Resources of the United States, 2017: a technical document supporting the Forest service 2020 RPA Assessment," U.S. Forest Services, accessed August 19, 2021, https://www.fs.usda.gov/treesearch/pubs/57903.

13. "State Of the World's Forests"

14. "Most Unwanted: Invasive Insects in U.S. Forests," Project Learning Tree, accessed December 11, 2021, https://www.plt.org/educator-tips/invasive-insects-forests/.

15. "State Of the World's Forests"

16. "Combined bark beetle outbreaks and wildfire spell uncertain future for forests," CU Boulder Today, accessed August 19, 2021, https://www.colorado.edu/today/2021/02/08/combined-bark-beetle-outbreaks-and-wildfire-spell-uncertain-future-forests.

17. "Are Bark Beetles Further Aggravating Wildfires in California?" Earth Island Journal, last modified March 25, 2021, https://www.earthisland.org/journal/index.php/articles/entry/are-bark-beetles-are-further-aggravating-wildfires-in-california.

18. "Small Pests, Big Problems: The Global Spread of Bark Beetles," Yale Environment 360, last modified September 21, 2017, https://e360.yale.edu/features/small-pests-big-problems-the-global-spread-of-bark-beetles.

19. "Fire and Bark Beetles – Resources from the NRFS website," Northern Rockies Fire Science Network, last accessed August 19, 2021, https://www.nrfirescience.org/fire-and-bark-beetles-resources-nrfsn-website.

20. F.C. Craighead, et al., "Control Work Against Bark Beetles in Western Forests and Appraisal of its Results," *Journal of Forestry* 29, no. 7 (1931), https://academic.oup.com/jof/article-abstract/29/7/1001/4571821?redirectedFrom=fulltext.

21. José F. Negròn, et al., "US Forest Service Bark Beetle Research in the Western United States: Looking Toward the Future," *Journal of Forestry* 106, no. 6 (2008), https://academic.oup.com/jof/article/106/6/325/4598840?-searchresult=1.

22. "State Of the World's Forests"

23. "Forest Fire/Wildfire Protection," Colorado Firecamp, last modified February 14, 2005, https://www.coloradofirecamp.com/congressional_research/forest-fire-wildfire-protection.htm.

24. "Forest Fire/Wildfire Protection"

25. "Forest Management," National Association of State Foresters, accessed August 22, 2021, https://www.stateforesters.org/where-we-stand/forest-management/.

26. "Federal Wildfire Management: Ten-Year Funding Trends and Issues (FY2011-FY2020)," Congressional Research Service, last modified October 28, 2020, https://crsreports.congress.gov/product/pdf/R/R46583.

ENDNOTES

27. "Forest Service Wildland Fire Suppression Costs Exceed $2 Billion," U.S. Department of Agriculture, last modified September 14, 2017, https://www.usda.gov/media/press-releases/2017/09/14/forest-service-wildland-fire-suppression-costs-exceed-2-billion#:~:text=%E2%80%9CForest%20Service%20spending%20on%20fire,forest%20management%2C%E2%80%9D%20Perdue%20said.

28. "2020 National Prescribed Fire Use Report," Coalition of Prescribed Fire Councils, Inc., accessed August 19, 2021, http://www.stateforesters.org/wp-content/uploads/2020/12/2020-Prescribed-Fire-Use-Report.pdf.

29. "Timber Harvesting."

30. "Toward Shared Stewardship Across Landscapes: An Outcome-Based Investment Strategy," United States Department of Agriculture, https://www.fs.usda.gov/sites/default/files/toward-shared-stewardship.pdf.

31. "O&C and Monumentizing: Dueling Rulings," The Smokey Wire: National Forest News and Views, last modified January 30, 2020, https://forestpolicypub.com/2020/01/30/oc-and-monumentizing-dueling-rulings/.

32. "Timber Harvesting."

33. "The Good Neighbor Authority," Congressional Research Service, last modified October 5, 2020, https://crsreports.congress.gov/product/pdf/IF/IF11658/3#:~:text=In%20its%20FY2021%20budget%20justification,and%2018%20for%20fire%20management.

34. "Society of American Foresters, Breakfast With the Chief," Forest Service and US Department of Agriculture, last modified November 18, 2017, https://www.fs.usda.gov/speeches/five-priorities-forest-service.

35. "Toward Shared Stewardship"

36. "Managing the Land," Forest Service and US Department of Agriculture, accessed August 19, 2021, https://www.fs.usda.gov/managing-land.

37. "USDA Forest Service Tribal Relations Strategic Plan Fiscal Year 2019-2022," United States Department of Agriculture, last modified December 2018, https://www.fs.fed.us/spf/tribalrelations/documents/plan/USDA-FS-TribalRelationsStrategicPlanFY2019-2022.pdf.

38. "Oregon-Washington Stewardship Contracting," US Department of the Interior Bureau of Land Management, last accessed August 19, 2021, https://www.blm.

gov/programs/natural-resources/forests-and-woodlands/stewardship-contracting/oregon.

39. Larry Mason et al., "Listening and Learning from Traditional Knowledge and Western Science: A Dialogue on Contemporary Challenges of Forest Health and Wildfire," *Journal of Forestry* 110, no. 4 (2012), accessed August 21, 2021, https://academic.oup.com/jof/article/110/4/187/4599525.

40. "Forestry Assessment," Intertribal Timber Council, accessed August 21, 2021, https://www.itcnet.org/issues_projects/issues_2/forest_management/assessment.html.

41. "Scenic Vista Management Plan for Yosemite National Park – Environmental Assessment, July 2010," National Park Service, US Department of the Interior, accessed August 21, 2021, https://www.nps.gov/yose/learn/management/upload/SVMP_YOSE_EA.pdf.

42. "Fire in Yosemite," National Park Service, last modified October 6, 2020, https://www.nps.gov/yose/learn/nature/wildlandfire.htm.

43. "Reasserting Tribal Forest Management Under Good Neighbor Authority," The Regulatory Review, last modified December 7, 2020, https://www.theregreview.org/2020/12/07/harris-reasserting-tribal-forest-management-good-neighbor-authority/.

44. "2018 Farm Bill Primer: Support for Indian Tribes," Congressional Research Service, last modified August 12, 2019, https://fas.org/sgp/crs/misc/IF11287.pdf#page=1.

45. Ronald L. Trosper, "Indigenous influence on forest management on the Menominee Indian Reservation," *Forest Ecology and Management* 249 (2007), https://courses.washington.edu/dtsclass/TEK-Menominee.pdf.

46. "Native American Forestry Combines Traditional Wisdom with Modern Science," Solutions Online, last modified February 22, 2016, https://thesolutionsjournal.com/2016/02/22/native-american-forestry-combines-traditional-wisdom-with-modern-science/.

47. "Forestry Assessment"

48. "Success in the Sierra: French Meadows Partnership doubles pace in the second season of work," Gold Country Media, last modified December 15, 2020, https://goldcountrymedia.com/news/187494/success-in-the-sierra-french-meadows-partnership-doubles-pace-in-second-season-of-work/.

ENDNOTES

49. "Restoring Forests through Partnership: Lessons Learned from the French Meadows Project," The Nature Conservancy, accessed August 21, 2021, https://www.scienceforconservation.org/assets/downloads/FrenchMeadows-Lessons_2019.pdf

50. "Success in the Sierra."

51. "French Meadows Forest Restoration Project," The Nature Conservancy, accessed August 21, 2021, https://www.nature.org/content/dam/tnc/nature/en/photos/FMPFactsheet2-27-19final.pdf.

52. "Tree Cities of the World," Food and Agriculture Organization of the United Nations and Arbor Day Foundation, accessed August 21, 2021, https://treecitiesoftheworld.org/.

53. "Who We Are," National Association of State Foresters, accessed August 21, 2021, https://www.stateforesters.org/who-we-are/.

54. "New York Declaration of Forests – Goal Assessments," New York Declaration on Forests, accessed August 21, 2021, https://forestdeclaration.org/goals.

55. "State Of the World's Forests"

56. Andrew C. Scott, et al., "The diversification of Paleozoic fire systems and fluctuations in atmospheric oxygen concentration," *Proceedings of the National Academy of Sciences of the United States of America* 103, no 29 (2006), https://www.ncbi.nlm.nih.gov/pmc/articles/PMC1544139/.

57. "Forest Fire/Wildfire Protection," Congressional Research Service, last modified March 7, 2012, https://fas.org/sgp/crs/misc/RL30755.pdf.

58. "WILDLAND FIRE—Federal Agencies' Efforts to Reduce Wildland Fuels and Lower Risk to Communities and Ecosystems," United States Government Accountability Office, last modified December 2019, https://www.gao.gov/assets/gao-20-52.pdf.

59. "A Call To End The Destruction Of Communities By Wildfire," National Fire Protection Association, accessed August 21, 2021, https://www.nfpa.org/-/media/Files/About-NFPA/Wildfire-policy/WildfirePolicy4.ashx.

60. "National Forest System Management: Overview, Appropriations, and Issues for Congress," Congressional Research Service, last modified September 5, 2019, https://fas.org/sgp/crs/misc/R43872.pdf.

61. "WILDLAND FIRE"

ENDNOTES

62. "Forest Fire/Wildfire Protection"

63. "WILDLAND FIRE"

64. "Forest Fire/Wildfire Protection"

65. Elizabeth L. Kalies, et al., "Tamm Review: Are fuel treatments effective at achieving ecological and social objectives? A systemic review," *Forest Ecology and Management* 375 (2016), accessed August 21, 2021, https://www.sciencedirect.com/science/article/abs/pii/S0378112716302626.

66. "WILDLAND FIRE"

67. "Defining Success for the Wildfire Funding Fix," Center for American Progress, last modified June 13, 2018, https://www.americanprogress.org/issues/green/reports/2018/06/13/451901/defining-success-wildfire-funding-fix/.

68. "Fires, Forests and the Future: A Crisis Raging Out of Control?" World Wide Fund For Nature and the Boston Consulting Group, accessed August 21, 2021, https://c402277.ssl.cf1.rackcdn.com/publications/1369/files/original/wwf_fires_forests_and_the_future_report.pdf?1598538474.

69. "The Rising Cost of Wildfire Operations: Effects on the Forest Service's Non-Fire Work," United States Department of Agriculture, last modified August 4, 2015, https://www.fs.usda.gov/sites/default/files/2015-Rising-Cost-Wildfire-Operations.pdf.

70. "WILDLAND FIRE"

71. "What is the WUI?" US Fire Administration accessed August 21, 2021, https://www.usfa.fema.gov/wui/what-is-the-wui.html.

72. "The 2010 Wildland-Urban Interface of the Conterminous United States," United States Department of Agriculture, accessed August 21, 2021, https://www.fs.fed.us/nrs/pubs/rmap/rmap_nrs8.pdf.

73. "WILDLAND FIRE"

74. "National Interagency Coordination Center—Wildland Fire Summary and Statistics Annual Report 2021," National Interagency Coordination Center, accessed May 31, 2022, https://www.predictiveservices.nifc.gov/intelligence/2021_statssumm/annual_report_2021.pdf.

75. "The Federal Land Management Agencies," Congressional Research Service, last modified February 16, 2021, https://fas.org/sgp/crs/misc/IF10585.pdf.

ENDNOTES

76. "WILDLAND FIRE"

77. "A Call to End the Destruction"

78. "US seeks to bolster firefighter ranks as wildfires increase," Associated Press News, last modified June 25, 2021, https://apnews.com/article/joe-biden-fires-business-environment-and-nature-government-and-politics-a40f55e3b8fa22e46748352e9aa4b443.

79. "Forest Service document reports 25% of hotshot crews can't meet required standards," Wildfire Today, last modified June 26, 2021, https://wildfiretoday.com/2021/06/26/forest-service-document-reports-25-of-hotshot-crews-cant-meet-required-standards/.

80. "WILDLAND FIRE"

81. "National Forest System Management"

82. "Evergreen. The Magazine of the Evergreen Foundation. Spring 2014," Intertribal Timber Council, accessed August 22, 2021, https://www.itcnet.org/file_download/eafa3eef-1775-4c09-b0a4-b961ee5d2277.

83. "2020 Best Places to Work in the Federal Government rankings," Partnership for Public Service, accessed December 11, 2021, https://bestplacestowork.org/rankings/?view=overall&size=large&category=leadership&.

84. "Trump signs Cantwell bill requiring new wildfire technology, smoke forecasts," The Associated Press, last modified March 21, 2019, https://apnews.com/article/78a82924594143f88a02a9193bf39554.

85. "National Environmental Policy Act Review Process," the United States Environmental Protection Agency, accessed August 22, 2021, https://www.epa.gov/nepa/national-environmental-policy-act-review-process.

86. "S.2882 – Wildfire Defense Act," Congress.Gov, accessed August 22, 2021, https://www.congress.gov/bill/116th-congress/senate-bill/2882.

87. "S. 4625 (116th): National Prescribed Fire Act of 2020," GovTrack, last modified September 17, 2002, https://www.govtrack.us/congress/bills/116/s4625/text.

88. "S. 1734: National Prescribed Fire Act of 2021," GovTrack, last modified May 20, 2021, https://www.govtrack.us/congress/bills/117/s1734.

89. "Mitt Romney calls for US commission to deal with wildfires," The Associated Press, last modified October 15, 2020, https://apnews.com/article/climate-climate-change-utah-legislation-wildfires-49707592157f84f9db36e9782e9c206a.

90. "Romney Joins Utah Officials to Unveil Plan to Improve Wildfire Policy," Mitt Romney U.S. Senator-Utah, last modified October 15, 2020, https://www.romney.senate.gov/romney-joins-utah-officials-unveil-plan-improve-wildfire-policy.

91. "Pending legislation related to wildland fire," Wildfire Today, last modified October 20, 2020, https://wildfiretoday.com/2020/10/20/pending-legislation-related-to-wildland-fire/.

92. "Congressmen LaMalfa, Panetta Introduce the Bipartisan, Bicameral Emergency Wildfire, and Public Safety Act," Congressman Doug LaMalfa, last modified August 7, 2020, https://lamalfa.house.gov/media-center/press-releases/congressmen-lamalfa-panetta-introduce-the-bipartisan-bicameral-emergency.

93. "Pending legislation"

94. "S.4431 – Emergency Wildfire and Public Safety Act of 2020," Congress.Gov, accessed August 22, 2021, https://www.congress.gov/bill/116th-congress/senate-bill/4431.

95. "H.R.7978 – Emergency Wildfire and Public Safety Act of 2020," Congress.Gov, accessed August 22, 2021, https://www.congress.gov/bill/116th-congress/house-bill/7978?r=4&s=1.

96. "H.R.939 – CARR Act," Congress.Gov, accessed August 22, 2021, https://www.congress.gov/bill/117th-congress/house-bill/939?r=22&s=1.

97. "H.R.1162 – 21st Century Conservation Corps Act," Congress.Gov, accessed August 22, 2021, https://www.congress.gov/bill/117th-congress/house-bill/1162.

98. "Lawmakers Propose Reforestation Expansion," The Nature Conservancy, last modified March 18, 2021, https://www.nature.org/en-us/newsroom/replant-act-reintroduced-house/.

99. "Lawmakers Introduce Bipartisan, Bicameral Legislation to Plant 1.2 Billion Trees on National Forests," United States Senate Committee on Agriculture, Nutrition, & Forestry, last modified March 18, 2021, https://www.agriculture.senate.gov/newsroom/dem/press/release/lawmakers-introduce-bipartisan-bicameral-legislation-to-plant-12-billion-trees-on-national-forests.

100. "H.R.3684 – Infrastructure Investment and Jobs Act," Congress.Gov, accessed December 11, 2021, https://www.congress.gov/bill/117th-congress/house-bill/3684/text.

101. "National Environmental Policy Act"

102. "Northern spotted owl's Endangered Species Act status will remain unchanged," Oregon Public Broadcasting, last modified December 15, 2020, https://www.opb.org/article/2020/12/15/northern-spotted-owl-endangered-species-act/.

103. "Recent and projected future wildfire trends across the ranges of three spotted owl subspecies under climate change," U.S. Forest Service, accessed August 22, 2021, https://www.fs.usda.gov/treesearch/pubs/57881.

104. "Northwest Forest Plan," Oregon Wild, accessed August 22, 2021, https://oregonwild.org/forests/forest-protection-and-restoration/nwfp.

105. Derek E. Lee, "Spotted Owls and forest fire: a systematic review and meta-analysis of the evidence," *Ecosphere* 97, no. 7 (2018), https://esajournals.onlinelibrary.wiley.com/doi/10.1002/ecs2.2354.

106. "California Knew the Carr Wildfire Would Happen. It Failed to Prevent It," ProPublica, last modified December 18, 2018, https://www.propublica.org/article/california-carr-wildfire-failed-to-prevent-it.

CHAPTER TEN

1. "Secretary-General's address at Columbia University: "The State of the Planet"," United Nations, accessed November 14, 2021, https://www.un.org/sg/en/content/sg/speeches/2020-12-02/address-columbia-university-the-state-of-the-planet.

2. Yude Pan, et al, "The Structure, Distribution, and Biomass of the World's Forests," *Annual Review of Ecology, Evolution, and Systematics* 44 (2013), https://www.annualreviews.org/doi/abs/10.1146/annurev-ecolsys-110512-135914.

3. "Forest Inventory and Analysis," USDA Forest Service, accessed October 7, 2021, https://www.fia.fs.fed.us/.

4. "Overview of Greenhouse Gases," United States Environmental Protection Agency, accessed October 7, 2021, https://www.epa.gov/ghgemissions/overview-greenhouse-gases.

5. T.A. Boden, et al, "Global, Regional, and National Fossil-Fuel CO2 Emissions (1751-2014)," Carbon Dioxide Information Analysis Center (CDIAC), Oak Ridge National Laboratory (ORNL), Oak Ridge, TN (United States), https://data.ess-dive.lbl.gov/view/doi:10.3334/CDIAC/00001_V2017.

6. "Change 2014: Mitigation of Climate Change," The Intergovernmental Panel on Climate Change, accessed October 7, 2021, https://www.ipcc.ch/report/ar5/wg3/.

7. "Overview of Greenhouse Gases."

8. "NOAA index tracks how greenhouse gas pollution amplified global warming in 2020," NOAA Research News, last modified May 24, 2021, https://research.noaa.gov/article/ArtMID/587/ArticleID/2759/NOAA-index-tracks-how-greenhouse-gas-pollution-amplified-global-warming-in-2020.

9. "Climate Change 2013: The Physical Science Basis," The Intergovernmental Panel on Climate Change, accessed October 7, 2021, https://www.ipcc.ch/report/ar5/wg1/.

10. "Voluntary carbon markets: how they work, how they're priced and who's involved," S&P Global, last modified Jun 10, 2021, https://www.spglobal.com/platts/en/market-insights/blogs/energy-transition/061021-voluntary-carbon-markets-pricing-participants-trading-corsia-credits.

11. "Taskforce on Scaling Voluntary Carbon Markets," The Institute of International Finance, accessed October 7, 2021, https://www.iif.com/tsvcm.

12. "FAQ: Forest Carbon Projects," The Climate Trust, last modified February 8, 2018, https://climatetrust.org/forest-carbon-projects-faq/.

13. "REDD+," United Nations Climate Change, accessed November 14, 2021, https://unfccc.int/topics/land-use/workstreams/reddplus.

14. "Ecosystem Marketplace's State of the Voluntary Carbon Markets 2021," Ecosystem Marketplace, accessed October 7, 2021, https://www.ecosystemmarketplace.com/publications/state-of-the-voluntary-carbon-markets-2021/.

15. "A blueprint for scaling voluntary carbon markets to meet the climate challenge," McKinsey Sustainability, last modified January 29, 2021, https://www.mckinsey.com/business-functions/sustainability/our-insights/a-blueprint-for-scaling-voluntary-carbon-markets-to-meet-the-climate-challenge.

16. "Taskforce on Scaling Voluntary Carbon Markets"

ENDNOTES

17. "Wall Street's Favorite Climate Solution Is Mired in Disagreements," Bloomberg Green, last modified June 2, 2021, https://www.bloomberg.com/news/features/2021-06-02/carbon-offsets-new-100-billion-market-faces-disputes-over-trading-rules.

18. "BP, the Colville Reservation Tribes and Finite Carbon Reach Milestone on West Coast Carbon Offset Project," BP, last modified December 15, 2016, https://www.bp.com/en_us/united-states/home/news/press-releases/bp-colville-reservation-tribes-and-finite-carbon-reach-milestone.html.

19. "Peru and Switzerland sign 'world first' carbon offset deal under Paris Agreement," Climate Home News, last modified on October 21, 2020, https://www.climatechangenews.com/2020/10/21/peru-switzerland-sign-world-first-carbon-offset-deal-paris-agreement/.

20. "United Nations Climate Change Reporting Requirements," United Nations, accessed November 26, 2021, https://unfccc.int/process-and-meetings/transparency-and-reporting/reporting-and-review-under-the-convention/greenhouse-gas-inventories-annex-i-parties/reporting-requirements.

21. "Inventory of U.S. Greenhouse Gas Emissions and Sinks," United States Environmental Protection Agency, accessed October 7, 2021, https://www.epa.gov/sites/default/files/2021-04/documents/us-ghg-inventory-2021-main-text.pdf?VersionId=wEy8wQuGrWS8Ef_hSLXHy1kYwKs4.ZaU.

22. "Census Bureau Population," United States Census Bureau, accessed October 7, 2021, https://www.census.gov/topics/population.html.

23. "What Is Carbon Capture and Storage," Climate Council, accessed October 7, 2021, https://www.climatecouncil.org.au/resources/what-is-carbon-capture-and-storage/.

CHAPTER ELEVEN

1. "*Autumn 1942 (Age 68)*," International Churchill Society, accessed August 5, 2021, https://winstonchurchill.org/the-life-of-churchill/war-leader/1940-1942/autumn-1942-age-68/.

2. "Wildland Urban Interface (WUI)," U.S. Fire Administration, last modified April 6, 2022, https://www.usfa.fema.gov/wui/.

ENDNOTES

3. "Flammable Planet: Wildfires and the Social Cost of Carbon," The Cost of Carbon Project, last modified September, 2014, https://costofcarbon.org/files/Flammable_Planet__Wildfires_and_Social_Cost_of_Carbon.pdf.

4. "Wildland Fire: Federal Agencies' Efforts to Reduce Wildland Fuels and Lower Risk to Communities and Ecosystems," U.S. Government Accountability Office, last modified December 19, 2019, https://www.gao.gov/products/gao 20 52.

5. "Hobson's choice," Wikipedia, accessed April 9, 2022, https://en.wikipedia.org/wiki/Hobson%27s_choice.

6. "Bark Beetles and Climate Change in the United States," USDA FS Climate Change Resource Center, last accessed February 2, 2022, https://www.fs.usda.gov/ccrc/topics/bark-beetles-and-climate-change-united-states.

7. "Grigory Potemik," Encyclopaedia Britannica, last accessed February 21, 2022, https://www.britannica.com/biography/Grigory-Potemkin.

8. "Observations on a Possible Move of the Forest Service into the Department of the Interior," United States Government Accountability Office, last modified February 2009, https://www.gao.gov/assets/gao-09-223.pdf.

9. "Americans Are Moving Closer to Nature, and to Fire Danger," The New York Times, last modified November 15, 2018, https://www.nytimes.com/2018/11/15/climate/california-fires-wildland-urban-interface.html.

10. "Land Use Planning Approaches in the Wildland-Urban Interface: An analysis of four western states: California, Colorado, Montana, and Washington," Community Wildfire Planning Center, last modified February 2021, https://www.communitywildfire.org/wp-content/uploads/2021/02/CWPC_Land-Use-WUI-Report_Final_2021.pdf.

11. Jennifer K. Balch, et al., "Human-started wildfires expand the fire niche across the United States," *Biological Sciences* 114, no. 11 (2017), accessed April 9, 2022, https://www.pnas.org/doi/10.1073/pnas.1617394114.

12. "Aerial firefighting resources in Europe," AirMed&Rescue, last modified April 13, 2021, https://www.airmedandrescue.com/latest/long-read/aerial-firefighting-resources-europe.

13. "Statement of Victoria Christiansen, Chief of the USDA Forest Service Before the House Committee on Appropriations Subcommittee on Interior, Environment, and Related Agencies Concerning the President's Fiscal Year 2022

ENDNOTES

Proposed Budget," USDA Forest Service, last modified April 15, 2021, https://docs.house.gov/meetings/AP/AP06/20210415/111425/HHRG-117-AP06-WstateChristiansenV-20210415.pdf.

14. "The Secret War Over Pentagon Aid in Fighting Wildfires," The New York Times, last modified September 27, 2021, https://www.nytimes.com/2021/09/27/science/wildfires-military-satellites.html.

15. "Wayne Gretzky Quotes," Goodreads, Inc., accessed April 9, 2022, https://www.goodreads.com/author/quotes/240132.Wayne_Gretzky.

16. "Suppression Costs," National Interagency Fire Center, accessed April 9, 2022, https://www.nifc.gov/fire-information/statistics/suppression-costs.

17. "About Wildfires," The Ad Council, accessed April 9, 2022, https://smokeybear.com/en/about-wildland-fire.

18. "The Paris Agreement," United Nations Framework Convention on Climate Change, last accessed April 9, 2022, https://unfccc.int/process-and-meetings/the-paris-agreement/the-paris-agreement.

19. "Outcomes of the Glasgow Climate Change Conference – Advance Unedited Versions (AUVs) and list of submissions from the sessions in Glasgow," United Nations Framework Convention on Climate Change, accessed April 9, 2022, https://unfccc.int/process-and-meetings/conferences/glasgow-climate-change-conference-october-november-2021/outcomes-of-the-glasgow-climate-change-conference.

20. "Carbon offsets are growing fast, but climate benefits remain murky," The Christian Science Monitor, last modified September 24, 2021, https://www.csmonitor.com/Environment/2021/0924/Carbon-offsets-are-growing-fast-but-climate-benefits-remain-murky.

21. "Carbon," U.S Department of Agriculture Forest Service, accessed April 9, 2022, https://www.fs.usda.gov/managing-land/sc/carbon.

22. "Lao Tzu Quotes," BrianyQuote, accessed April 9, 2022, https://www.brainyquote.com/quotes/lao_tzu_151993.

23. "Infrastructure Investment and Jobs Act of 2021, Public Law 117-58," United Stated Government Publishing Office, last modified November 15, 2021, https://www.congress.gov/117/plaws/publ58/PLAW-117publ58.pdf.

24. "USDA Forest Service Tribal Relations Strategic Plan Fiscal Year 2019-2022," United States Department of Agriculture, last modified December 2018, https://www.fs.usda.gov/sites/default/files/fs_media/fs_document/trsp_handbook_b_final_508_compliance_smallsize.pdf.

25. "Science supporting shared stewardship at the Rocky Mountain Research Station," U.S. Forest Service, last modified July 18, 2019, https://www.fs.usda.gov/rmrs/documents-and-media/science-supporting-shared-stewardship-rocky-mountain-research-station.

26. "The Venado Declaration," Associated Press, accessed April 9, 2022, https://www.documentcloud.org/documents/21100767-venado-declaration.

27. "Opportunity Zones," IRS, last modified November 10, 2021, https://www.irs.gov/credits-deductions/businesses/opportunity-zones#:~:text=Opportunity%20Zones%20were%20created%20under,zones%20through%20Qualified%20Opportunity%20Funds.

28. "Companies Plan to Pour Even More Cash Into Buybacks, Dividends in 2022," The Wall Street Journal, last modified December 23, 2021, https://www.wsj.com/articles/companies-plan-to-pour-even-more-cash-into-buybacks-dividends-in-2022-11640169002?mod=Searchresults_pos7&page=1.

29. "Monthly Budget Review: Summary for Fiscal Year 2021," Congressional Budget Office, last modified November 8, 2021, https://www.cbo.gov/publication/57539.

30. "U.S. National Debt Tops $30 Trillion as Borrowing Surged Amid Pandemic," The New York Times, last modified by February 1, 2022, https://www.nytimes.com/2022/02/01/us/politics/national-debt-30-trillion.html.

31. Other FPZ governance structure issues we suggest include:

    - No person or corporate sponsor shall serve on more than one FPZ council. No council member can acquire any carbon credit offsets from the FPZ they represent.

    - Each council shall agree to a rotation process for the chair chosen from one of the council members.

    - Provide federal and state veto power over decisions proposed by individual FPZ councils with rights granted to the councils for redress through the media and courts when issues arise with bureaucratic decision.

- Establish a sunset period of fifteen years for each FPZ. At the conclusion of the term, compare each zone's effectiveness and best practices for meeting its mission.

- Based on the results of this assessment, revise the mission statements and forest protection regions as necessary to incorporate the best forest management practices and regional governance successes from the initial incubator FPZ results. Use any remaining surpluses from the initial FPZ to fund new forest protection regions.

- Retain governance control over wildfire suppression and forest management activities in their respective zones, including conservation, restoration, and sustainable use of terrestrial and inland freshwater systems, wetlands, mountains, and drylands.

- Maintain immediate access to government-owned aerial, bulldozer, personnel, and firefighting assets.

- Establish best practices to put out wildfires immediately.

- Work with existing GACCs and the National Interagency Coordination Center/NIFC to coordinate and preposition firefighting assets through the use of advanced predictive tools for quicker response times. The goal is to reach and establish meaningful perimeters around wildfires to extinguish them within the first two days of an outbreak.

- Organize each FPZ as a mutual-benefit, not-for-profit corporation. Independent third-party firms with appropriate expertise will conduct annual required audits of the zone's financial statements and triennial audits of its forest stewardship activities. These reports will be made public and permanently maintained on websites for each FPZ to record achievements permanently.

- Coordinate, as needed, with other FPZs to share assets, research, and ideas.

CONCLUSION

1. "Henry David Thoreau Quotes," Brainy Quote, accessed August 5, 2021, https://www.brainyquote.com/authors/henry-david-thoreau-quotes.

2. "Our National Parks—Chapter 2," Sierra Club, accessed February 13, 2022, https://vault.sierraclub.org/john_muir_exhibit/writings/our_national_parks/chapter_2.aspx.

ENDNOTES

3. "10 John Muir Quotes That'll Inspire You to Explore America's Great Outdoors," U.S. Department of the Interior, last modified April 13, 2016, https://www.doi.gov/blog/10-john-muir-quotes-that-ll-inspire-you-explore-america-s-great-outdoors.

BIBLIOGRAPHY

Excludes Peer-Reviewed Research Reports, Journal Articles, Newspaper Reports, Training Materials, and Internet Website Research Results Set Forth in the References to each Individual Chapter

BACKGROUND BOOKS

1. Anderson, M. Cat. *Tending the Wild: Native American Knowledge and the Management of California's Natural Resources*, 2005.

2. Bartz, Kelsey. Record Wildfires Push 2018 Disaster Costs to $91 Billion. Center for Climate and Energy Solutions, 2019.

3. Carle, David. *Introduction to Fire in California*, 2021.

4. Ferguson, Gary. *Land on Fire: The New Reality of Wildfire in the West*, 2017.

5. Fletcher, William D. and Smith, Craig B. *Reaching Net Zero: What it Takes to Solve the Global Climate Crisis*, 2020.

6. Gee, Alastair and Anguiano, Dani. *Fire in Paradise: An American Tragedy*, 2020.

7. Johnson, Bridget. "World's Worst Wildfires." ThoughtCo.com, 2019.

8. Johnson, Lizzie. *Paradise: One Town's Struggle to Survive an American Wildfire*, 2021.

9. Matthews, Daniel. *Trees in Trouble: Wildfires, Infestations, and Climate Change*, 2020.

10. Nix, Steve. *The Origin of Wildfires and How They Are Caused*, 2018.

11. Palley, Stuart. *Terra Flamma: Wildfires at Night*, 2018.

12. Petersen, James D. "First, Put Out the Fire!" Evergreen, 2020.

13. Pyne, Stephen J. *Between Two Fires: A Fire History of Contemporary America*, 2015.

14. Pyne, Stephen J. *Fire in America: A Cultural History of Wildland and Rural Fire*, 1997.

15. Ribe, Tom. *Inferno by Committee: A History of the Cerro Grande (Las Alamos) Fire America's Worst Prescribed Fire Disaster*, 2010.

16. Scott, Andrew C. *Burning Planet: The Story of Fire Through Time*, 2018.

17. Stewart, Omer C. *Forgotten Fires, Native Americans and the Transient Wilderness*, 2002.

18. Thomas, Douglas S.; Butry, David T.; Gilbert, Stanley W.; Webb, David H.; and Fung, Juan F. "The Costs and Losses of Wildfires: A Literature Survey." *Environmental Science*, 2017.

FOREST MANAGEMENT

1. Ager, Alan A.; Day, Michelle A.; Waltz, Amy; Nigrelli, Mark; Vogler, Kevin C.; and Lata, Mary. "Balancing Ecological and Economic Objectives in Restoration of Fire-Adapted Forests: Case Study from the Four Forest Restoration Initiative." U.S. Department of Agriculture, 2021.

2. Bureau of Indian Affairs. *Indian Forest Management Handbook*, 53 IAM 2-H, March 19, 2009.

3. California Department of Forestry and Fire Protection (Cal Fire). *Community Wildfire Prevention and Mitigation Report*, 2019.

4. Texas A & M Forest Service. *Community Wildfire Protection Plan Guide*. Wildland Urban Interface Program, 2012.

5. Evergreen Foundation. Forestry in Indian Country: Models of Sustainability for Our Nation's Forests? *Evergreen Magazine*, 2005.

6. Fire Executive Council. *Guidance for Implementation of Federal Wildland Fire Management Policy*, 2009.

7. Food and Agricultural Organization of the United Nations. *The State of the World's Forests*, 2020.

8. German Federal Agency for Nature Conservation. *Sustainable Forest Management in Germany: The Ecosystem Approach of the Biodiversity Convention Reconsidered Results of the R+D–Project*, 2002.

9. German Federal Ministry of Food and Agriculture. *German Forests—Forests for Nature and People*, 2021.

10. Graham, Russell T.; Battaglia, Mike A.; and Jain, Theresa B. *A Scenario-Based Assessment to Inform Sustainable Ponderosa Pine Timber Harvest on the Black Hills National Forest*. U.S. Department of Agriculture, 2021.

11. Hauger Lindstad, Berit. *A Comparative Study of Forestry in Finland, Norway, Sweden, and the United States, with Special Emphasis on Policy Measures for Nonindustrial Private Forests in Norway and the United States*. U.S. Department of Agriculture, 2002.

12. Hogland, John; Dunn, Christopher J.; and Johnston, James D. "21st Century Planning Techniques for Creating Fire-Resilient Forests in the American West." *Forests*, 2021.

13. Legislative Analyst's Office. *Improving California's Forest and Watershed Management*, 2018.

14. National Association of State Foresters website. www.stateforesters.org, 2021.

15. New York Declaration on Forests (NYDF). NYDF Progress Assessment reports and NYDF Platform publication, 2014-2021.

16. Recommendations of the (California) Governor's Forest Management Task Force, 2021.

17. State of Texas. Helping Hands Program, HB 680, 1997, 75th Legislative Session; Rural Volunteer Fire Department Assistance Program, HB 2604, 2001, 77th Legislative Session; Statewide Mutual Aid-SB 11, 2007, 80th Legislative Session.

18. The Nature Conservancy. Placer County Water Authority, Sierra Nevada Conservancy, County of Placer, American River Conservancy, University of California-Merced, *Restoring Forests through Partnership: Lessons Learned from the French Meadows Project*, 2019.

19. United States Department of the Interior, Bureau of Indian Affairs. *Forest Management Plan, Yakama Reservation*, September 2005.

20. USDA Forest Service. Forest Inventory and Analysis (FIA) Reports, 2021.

21. USDA Forest Service. Good Neighbor Authority Program, 2018.

22. USDA Forest Service, Shared Stewardship Program, 2018-2021.

23. USDA Forest Service. The Principal Laws Relating to USDA Forest Service State and Private Forestry Programs, 2011.

24. United States Government Publishing Office. Oversight Hearing before the Subcommittee on Public Lands and Environmental Regulation of the Committee on Natural Resources U.S. House of Representatives, 2013.

25. University of Oregon. *Confederated Tribes of Warm Springs Reservation Natural Hazard Mitigation Plan*, July 2016.

26. U.S. Senate Committee on Energy and Natural Resources. Testimony of John "Chris" Maisch, Alaska State Forester On Behalf of the National Association of State Foresters, 2019.

27. Yu, Anthony C.; Hernandez, Hector Lopez; Kim, Andrew H.; Stapleton, Lyndsay M.; Brand, Reuben J.; Mellor, Eric T.; et al. Wildfire Prevention Through Prophylactic Treatment of High-Risk Landscapes Using Viscoelastic Retardant Fluids. *Proceedings of the National Academy of Sciences*, 2019.

COSTS FOR WILDFIRE SUPPRESSION AND FOREST MANAGEMENT

1. Boston Consulting Group. *Fires, Forests, and the Future*, 2020.

2. Boston Consulting Group. *The Staggering Value of Forests and How to Save Them*, 2020.

3. Congressional Research Service. *Proposals to Merge the Forest Service and the Bureau of Land Management: Issues and Approaches*, 2008.

4. Cook, Philip S. and Becker, Dennis R. *State Funding for Wildfire Suppression in the Western U.S.* University of Idaho, 2017.

5. Evarts, Ben. *Fire Loss in The United States During 2018*. National Fire Protection Association, 2019.

6. Headwaters Economics. *The Full Community Costs of Wildfire*, 2018.

7. Insurance Information Institute. *Background on: Wildfires*, 2018.

8. Insurance Information Institute. *Facts + Statistics: Wildfires*, 2021.

BIBLIOGRAPHY

9. Munich RE, "Record Hurricane Season and Major Wildfires—the Natural Disaster Figures for 2020" (and similar report for 2018 and 2019), 2021.

10. Orange County (CA) Fire Authority. Fiscal Year 2019/20 Adopted Budget, 2019.

11. Swiss Re Institute. "Sigma No. 1/2021—National Catastrophes in 2020: Secondary Perils in the Spotlight but Don't Forget the Primary Peril Risks," 2021.

12. The Nature Conservancy and Willis Towers Watson. *Wildfire Resilience Insurance: Quantifying the Risk Reduction of Ecological Forestry with Insurance*, 2021.

13. The Pew Charitable Trusts. *How States Pay for Natural Disasters in the Era of Rising Costs*, 2020.

14. The Pew Charitable Trusts. *What We Don't Know About State Spending on Natural Disasters Could Cost Us*, 2018.

15. United States Department of Agriculture. *United States Forest Service, Budget Justification, Fiscal Year 2021 and Fiscal Year 2022*.

16. United States Department of Homeland Security, FEMA. *A Guide to the Disaster Declaration Process and Federal Disaster Relief*, 2021.

17. United States Department of Homeland Security. *Fiscal Year 2021 Report to Congress, Disaster Relief Fund: Monthly Report*, 2020.

18. United States Department of Interior. *Budget Justification, Fiscal Year 2021 and Fiscal Year 2022*.

19. United States Government Accountability Office (GAO). *Budgeting for Disasters - Approaches to Budgeting for Disasters in Selected States*, 2015.

20. United States Government Accountability Office (GAO). *Federal Land Management - Observations on a Possible Move of the Forest Service into the Department of Interior*, 2009.

WILDFIRE MANAGEMENT, LAND OWNERSHIP, AND RELATED STATISTICS

1. Glacier National Park. Flathead National Forest Fire Information, Fire Terminology, 2003.

2. Library of Congress. Congressional Research Service Reports (various, regarding annual wildfire statistic updates, wildfire suppression costs, federal land ownership, carbon data), 2019-2021.

3. National Interagency Coordination Center. *Wildland Fire, Summary and Statistics, Annual Report*, 2019 and 2020.

4. National Multi-Agency Coordinating Group. *National Multi-Agency Coordinating Group Operations Plan 2020*, 2020.

5. National Weather Service. *National Fire Weather Annual Operating Plan*, 2021.

6. United States Department of Agriculture. *Spatial Wildfire Occurrence Data for the United States*. Research Data Archive, 1992-2018 with online updates through 2020.

7. United States Departments of Interior and Agriculture. *Guidance for Implementation of Federal Wildland Fire Management Policy*, 2009.

8. United States Environmental Protection Agency. *Wildfire Smoke: A Guide for Public Health Officials*, 2019.

9. USDA Forest Service. *Memorandum of Understanding between USDA Forest Service, State and Private Forestry, and the Department of the Interior*, 2016.

10. USDA Forest Service. *Service-Wide Memorandum of Understanding between USDA Forest Service and DHS Federal Emergency Management Agency*, 2021.

CARBON MANAGEMENT

1. Apple, Inc. *Environmental Progress Report*, Fiscal Year 2020.

2. Congressional Research Service. *Aviation and Climate Change*, 2021.

3. Hawkins, Slayde. *Building Forest Carbon Projects, Legal Guidance—Legal and Contractual Aspects of Forest Carbon Projects*. Forest Trends, 2011.

4. Hawkins, Slayde; Nowlin, Michelle; Ribeiro, Daniel; Stoa, Ryan; Longest, Ryke; and Salzman, Jim. *Contracting for Forest Carbon: Elements of a Model Forest Carbon Purchase Agreement*, 2010.

5. International Energy Agency. *Global Energy Review 2021—Assessing the Effects of Economic Recoveries on Global Energy Demand and CO2 Emissions in 2021*, 2021.

6. Norwegian Ministry of Climate and Environment. *National Forestry Accounting Plan for Norway for the First Commitment Period 2021-2025*, 2019.

7. Smith, Gordon. *Forest Offset Projects on Federal Lands*. Climate Action Reserve, 2012.

8. Taskforce on Scaling Voluntary Carbon Markets. *Final Announcement of the Recommendation for the New Governance Body Composition*, January 2021.

9. *The Economist*. The Climate Issue, 2019.

10. United Nations. *Climate Change—United Nations Paris Agreement*, 2015.

11. United Nations. *Kyoto Protocol to the United Nations Framework Convention on Climate Change*, 1997.

12. United States Environmental Protection Agency. *Inventory of U.S. Greenhouse Gas Emissions and Sinks 1990-2019*, 2021.

13. Washington State Department of Natural Resources. *Carbon Sequestration Advisory Group Report*, 2020.

14. *Wildfires and Climate Change: California's Energy Future, A Report from Governor Newsom's Strike Task Force*, 2019.

15. World Bank Group. *State and Trends of Carbon Pricing 2020*, 2020.

AVIATION, MECHANICAL EQUIPMENT, AND WILDFIRE MANAGEMENT

1. Arizona Department of Forestry and Fire Management. *Yarnell Hill Fire, June 30, 2013, Serious Accident Investigation Report*, 2013.

2. California Department of Forestry and Fire Protection (Cal Fire). *Firefighting Aircraft Recognition Guide*, 2021.

3. California Department of Forestry and Fire Protection (Cal Fire). Wildfire Activity Statistics, 2018, 2019, 2020.

4. Department of the Interior Bureau of Land Management, National Park Service, United States Fish and Wildlife Service, Bureau of Indian Affairs, USDA Forest Service. *Interagency Standards for Fire and Fire Aviation Operations, NFES 272*, 2021.

5. Jaffe, Valerie and O'Brien, Stephen "Obie." *Mechanized Equipment for Fire and Fuel Operations*, 2009.

6. JDA Journal. "A Repeat Request for a New Firefighting Aircraft," 2018.

7. National Fire Protection Association. *U.S. Fire Department Profile—2018*, 2020.

8. National Interagency Coordination Center. *National Dispatch Standard Operating Guide for Contracted Resources*, 2018.

9. National Interagency Fire Center. *Northern Rockies Solicitation Plan for Competitive EERAS*, 2021.

10. National Wildfire Coordinating Group. *Gaining an Understanding of the National Fire Danger Rating System*, 2002.

11. National Wildfire Coordinating Group. *Incident Response Pocket Guide*, 2018.

12. National Wildfire Coordinating Group. *Master Interagency Agreement for National Wildfire Coordinating Group (NWCG) Share Funding among the BLM, BIA, FWS, and NPS of the United States Department of the Interior and the U.S. Forest Service (USFS) of the Department of Agriculture*, 2020.

13. National Wildfire Coordinating Group. *NWCG Airtanker Base Directory*, 2020.

14. National Wildfire Coordinating Group. *NWCG Standards for Wildland Fire Module Operations*, 2019.

15. National Wildfire Coordinating Group. *S-130, Firefighter Training*, 2020.

16. National Wildfire Coordinating Group. *Wildland Fire Incident Management Field Guide*, 2014.

17. Northern Rockies Coordination Center. "Contracting For Fire—Equipment Requirements," 2021.

18. Refai, Razim and Paskaluk, Stephen. *Evaporation Rates: Assessing the Evaporation Rates of Water, Foam, and Water-enhancers*. FP Innovations, 2021.

BIBLIOGRAPHY

19. United States General Accounting Office. *Report to the Chairman, Committee on Agriculture, House of Representatives, Forest Service-Weak Contracting Practices Increase Vulnerability to Fraud, Waste, and Abuse*, 1998.

20. USDA Forest Service Fire & Aviation Management and Department of Interior Office of Wildland Fire. *2014 Quadrennial Fire Review*, 2015.

21. USDA Forest Service. *How to Generate and Interpret Fire Characteristics Charts for Surface and Crown Fire Behavior*, 2011.

22. USDA Forest Service. *Standards for Airtanker Operations*, 2019.

23. USDA Forest Service National Office. *Exclusive Use Next Generation Large Airtanker Services 3.0*, 2017.

24. The Rand Corporation, Homeland Security and Defense Center. *Air Attack Against Wildfires—Understanding U.S. Forest Service Requirements for Large Aircraft*, 2012.

25. Waters, Keith L. and Fuller, Stephen S. *The Impact of Utilizing Aerial Tankers in Fighting Forest Fires*, 2020.

ACKNOWLEDGMENTS

To our families and close friends, who never let us forget the importance of the struggle to expose this subject matter, and who have supported us and pushed us to keep up the demanding work of writing this book.

To our many friends and colleagues who sent us articles about the historic amount and ferocity of wildfires that occurred between 2009 and 2021. They never let us forget the personal impact of their own experiences involving their actual and potential evacuations during these tragic events.

To the members of the USDA Forest Service, State Forestry Agencies, and other firefighting agencies who made the ultimate sacrifice, we extend our condolences to your families. Thank you for keeping so many safe while living through the horror of wildfires.

To the staff and retirees across the various offices of the USDA Forest Service in Washington, DC, Boise, ID; Fort Collins, CO; and various offices in the fourteen western, southwestern, and southern states who helped us understand the scope and direction of the exceptional efforts made to move the USDA Forest Service to the future. Our interaction with the National Interagency Fire Center staff was enlightening. The frustration levels must be enormous.

To the management and staff of the USDA Forest Service involved with the arduous task of tracking and recording individual wildfire events across the United States. The details of your work should be the foundation of planning for the future of putting out the fires.

To the offices of the governors of Texas, Utah, Arizona, and Montana, and the various state forestry offices who played a significant role in guiding our initial exposure and first steps into the business side of the forestry and timberland industries, thank you!

To the staff and the operating personnel in the maintenance, contracting, flight operations, and general management of the companies who work with the USDA Forest Service, we deeply appreciate the skills you acquired to keep yourselves safe and still provide your services in extremely dangerous environments.

To the many senior staffers of CAL FIRE who provided us with an education about the legal, aviation sector, training, and forestry operations across the state. You were invaluable in helping us understand the interactions and sense of urgency in responding to and dealing with the myriad issues and challenges involving wildfire urban interface.

To the college and university deans and their immediate staffs of the forestry departments who guided us to the proper in-state organizations that were invaluable in providing us a view of the financial, legal, operational, and local governmental aspects of fighting wildfires in their respective states.

To the many scholars involved in writing the countless papers we reviewed (too many to acknowledge) and depended upon in making observations that we hope will assist in making the USDA Forest Service more interactive with its many audiences.

To the broad spectrum of suppliers engaged with the USDA Forest Service and other local, state, and federal agencies, we acknowledge and appreciate the guidance and education provided to us for finding the right sources of information in those agencies. In some cases, those contacts were indeterminable.

To the contacts we made who were open, honest, and circumspect in their comments and who guided us in asking the questions that helped shape our writing about the agencies, their roles and missions, and our understanding of the complexities of fighting wildfires.

To the same people and supplier companies who have a history with senior staffs of specific agencies for their insights into the organizational construct of and interactions between the USDA Forest Service, the Bureau of Indian Affairs, the Bureau of Land Management, the Department of Interior and the people they serve.

To the various state adjutant generals and the U.S. Department of Defense for trying to reach agreement on shared responsibilities for firefighting and paying for the assets under Title 32 for fighting wildfires.

To the state foresters, counties, and communities charged with the care of our local, state and national forestlands. They should be credited with the enormous job in garnering ground, local-level support and education systems that fund and educate, train, and assist local fire departments' responses to potential wildfires. They are the creative group for finding ways to work with the many companies and sectors of the commercial world who support and assist the forestry and timberland industries.

To the aircraft manufacturers that provided countless answers to questions about structural issues, flight operations, built-in safety systems, flight training, and technical information related to aging aircraft, all of which still serve the USDA Forest Service and other agencies involved in aerial firefighting.

To the trade associations that built significant annual exhibitions to display new and developing equipment for their industries for their assistance providing information related to attracting industry partners and service providers to interact with all levels of their governmental partners, specifically, the defense, aviation, and agricultural industry.

To the U.S. forestry and wildfire fighting press that consistently advocate, educate, and develop a readership following the extraordinary events occurring in these growing markets.

To the many entrepreneurs, innovators, and companies who spend countless hours and resources in developing new products and services in the United States and internationally, that mostly have gone unrecognized.

To our colleague and good friend, Richard McAdoo, who provided valuable insight into many technical aspects concerning aircraft operations and cost metrics. To Dave's wife, Barbara, whose patience and understanding allowed him to write this book. And finally, to our colleagues Gerri Knilans, Laura Eckhardt, and Jamie Pagett of Trade Press Services, Iler Ganz of Baywind Services, John Downey, and Joseph McDonnell, all of whom have been unwavering in their assistance to help shape, edit, and chart our work. God bless you all!

ABOUT THE AUTHORS

 David L. Auchterlonie has more than 53 years of corporate management expertise. In addition to his 18 years with Price Waterhouse as a senior audit manager and as a financial officer for three public companies, he founded a corporate turnaround management firm that he headed for more than 30 years. Using his Certified Turnaround Management training and expertise, his firm provided hands-on management to more than 200 companies preserving more than 25,000 jobs and developed strategies to create long-term value for each client. David completed over 40 mergers and acquisitions transactions with valuations totaling in excess of $4.5 billion and raised above $2 billion of debt and equity. In addition, he successfully restructured more than $2 billion of private and public debt during his career. David's industry expertise includes aviation/aerospace, consumer products, food, transportation, distribution, healthcare, manufacturing, and technology, among others.

In 2008, the Turnaround Management Association recognized his turnaround and leadership accomplishments by inducting him into its Hall of Fame. David has published seven articles in peer-reviewed journals, two of which received awards from the League of American Communications Professionals. He has also been a featured speaker at more than 10 conferences, forums, and events.

 Jeffrey A. Lehman brings global expertise to public, private, and governmental clients with more than 50 years of executive experience. Working with Fortune 50 U.S. companies and governmental agencies, Jeff provides his vast know-how on multinational logistics, security, organizational complexities and sales management.

An expert in airline operations, Jeff conceptualized, launched, directed and staffed World Airways multi-continent cargo operation where he served as vice president. As vice president of Lockheed's Air Terminal company, Jeff headed the marketing department overseeing the conversion of an underutilized military airport into a dual use regional commercial-military airport. During that time, he worked with the Lt. Governor, state assembly and senate organizations as well as the New York State Congressional delegation and a politically appointed board of airport commissioners. He worked directly with multiple agencies of the U.S. Department of Transportation, including the administrators of the FAA and senior executives in various U.S. and foreign airlines. As the vice president of private industry affairs for the Airport Council International (formerly the Airport Operators Council International), Jeff was responsible for the design and implementation of programs that represented the collective interests of the private sector, airports, governments, subject matter experts and organizations in promoting excellence in advancing standards in the global airport and aviation industries. He has received recognition awards from the Secretary of Transportation and the FAA for spearheading the opening of significant airspace in the northeast corridor of the United States.

Jeff also served as senior vice president and general manager of Mid Pacific Airport Corporation (MPAC) where he was responsible for the transformation of the airline from a cargo charter carrier to a regional scheduled airline working with other air and integrated carriers across various market sectors. During his tenure, he was also involved with the selection of new aircraft and aircraft leasing, operations, customer service, route planning and pricing.

Additional engagements included serving as an international aviation consultant with McDonnell Douglas and Boeing Defense Systems on the development of the commercial variant of the C-17 aircraft and its unique capabilities to change management of global logistics as well as for use by commercial entities and various governmental agencies. His involvement led to deep engagements with the U.S. Departments of Defense, State, Homeland Security and the Government Printing Office, as well as the ministries of defense of the United Kingdom, Canadian Department of National Defense and Defense Australia.

He served on the boards of National Industrial Transportation League, Washington, DC, the George C. Marshall International Center of Leesburg, VA, The Corporate Angel Network, White Plains, NY and the Greater Washington Aviation Open a charity focused on providing transportation for cancer patients to and from treatment centers and for the Veterans Uplift Command.

Most recently, Jeff co-founded a boutique consulting firm, Crowbar Research Insights, LLC™ to perform studies to examine, analyze and improve the effectiveness of complex organizations.